# Publishing from your Doctoral Research

T0326108

Today's researchers ha...
disseminating their work, including traditional and digital publications, written articles, podcasts, and other media such as zines and graphic books. But how do they decide which output is right for them, and where to start? *Publishing from your Doctoral Research* provides methods and tools to help assess, identify, and adapt academic work for publication to support career aspirations.

Discussing what publication can achieve in career terms, this book:

- Explains how to audit doctoral research, and any associated materials, to assess which elements are best suited for publication
- Provides advice on how to determine what kind of publication is best suited to different types of research
- Discusses journal articles, books, self-publishing, online and social media options, and alternative methods of publishing
- Considers each type of publication in light of career aspirations

Each chapter includes practical examples, tailored to researchers interested in working in academia, industry or business, a clinical or practical career, or self-employment. Providing key strategies and insights to secure knowledge and success, *Publishing from your Doctoral Research* is the ideal guide for anyone looking to develop their career through publication within or outside academia.

**Janet Salmons** is an independent researcher, writer, and coach based in the United States.

**Helen Kara** is an independent researcher and writer based in the United Kingdom.

# Insider's Guides to Success in Academia

Series Editors:

Helen Kara
*Independent Researcher, UK, and*

Patricia Thomson
*The University of Nottingham, UK*

The *Insider's Guides to Success in Academia* address topics too small for a full-length book and too big to cover in a single chapter or article. These topics have often been the stuff of discussions on social media or of questions in our workshops. We designed this series to answer these questions and to provide practical support for doctoral and early-career researchers. It is geared to concerns that many people experience. Readers will find these books to be helpful and reassuring companions that provide advice and strategies to make sense of everyday life in the contemporary university. We have invited scholars with deep and specific expertise to write. Our writers use their research and professional experience to provide well-grounded strategies to particular situations. We have also asked writers to collaborate. Most of the books are produced by writers who live in different countries, or work in different disciplines, or both. While it is difficult for any book to cover all the diverse contexts in which potential readers live and work, the different perspectives and contexts of writers go some way to address this problem.

We understand that the use of the term 'academia' might be read as meaning the university, but we take a broader view. Pat does indeed work in a university but spent a long time working outside of one. Helen is an independent researcher and sometimes works with universities. Both of us understand academic – or scholarly – work as now conducted in a range of sites, from museums and the public sector to industry research and development laboratories. Academic work is also often undertaken by networks which bring together scholars in various virtual and real-world locations. All our writers understand that this is the case and use the term 'academic' in this wider sense.

These books are pocket sized so that they can be carried around and visited again and again. Most of the books have a mix of examples, stories, and exercises as well as explanation and advice. They are written in a collegial tone and from a position of care as well as knowledge.

Together with our writers, we hope that each book in the series can make a positive contribution to the work and life of readers, so that you too can become insiders in scholarship.

Helen Kara, PhD, FAcSS
Independent researcher
https://helenkara.com/
@DrHelenKara (Twitter/Insta)
Pat Thomson, PhD, PSM FAcSS FRSA
Professor of Education, The University of Nottingham
https://patthomson.net
@ThomsonPat

**Books in the Series:**

**Publishing from your Doctoral Research**
Create and Use a Publication Strategy
*Janet Salmons and Helen Kara*

# Publishing from your Doctoral Research

## Create and Use a Publication Strategy

## Janet Salmons and Helen Kara

Routledge
Taylor & Francis Group

LONDON AND NEW YORK

First published 2020
by Routledge
2 Park Square, Milton Park, Abingdon, Oxon, OX14 4RN

and by Routledge
52 Vanderbilt Avenue, New York, NY 10017

*Routledge is an imprint of the Taylor & Francis Group, an informa business*

*British Library Cataloguing-in-Publication Data*
A catalogue record for this book is available from the British Library

*Library of Congress Cataloging-in-Publication Data*
A catalog record has been requested for this book

ISBN: 978-1-138-33913-2 (hbk)
ISBN: 978-1-138-33914-9 (pbk)
ISBN: 978-0-429-44125-7 (ebk)

Typeset in Helvetica
by Cenveo® Publisher Services

Visit the eResources: www.routledge.com/9781138339149

Janet dedicates the book to Hannah, Zachary, Alex, Sam, and Oliver. Keep sharing your ideas and asking questions!

Helen dedicates the book to its readers. Yep – that means you.

# Contents

# Figures

# Tables

# Using this book

Students' questions 'What should I publish?' and 'How can I use my dissertation?' inspired Janet Salmons to create the publishing typology central to this book. The questions 'Must research-based writing be published in journal articles?' and 'Does academic writing have to be in *writing*?' inspired Helen Kara to look for creative options for disseminating research. We brought our ideas and experiences together to develop the approaches and resources offered here and on our website, www. path2publishing.com. Between the two of us, we have tried every type of publishing included in this book. We've worked with traditional publishers and experimented with DIY publishing. We understand that the expanding choices, and associated work to actually follow through and complete each project, can seem overwhelming. We know that by creating a strategy, it's possible to feel more in control of the process.

We hope this activity-based guide will encourage you to explore ways to advance in your career and make a difference by publishing your work. You will find the tools and motivation you need to create a multi-year publication strategy that includes both traditional and ground-breaking ways to get your research out into the world. Naturally, in a book of this length, we couldn't cover every possible publication type, so we will continue to add resource materials to our website.

We acknowledge that this book is oriented towards people writing in English and the Western expository style. We recognise how difficult that can be for people with a different first language or culture. Look for resources on www.path2publishing.com.

Together, we share a commitment to the inclusion of traditional and non-traditional scholars in the important conversations of our time. We believe that by welcoming wider participation in all kinds of publishing, we open the conversation to include voices we all need to hear. We look forward to hearing yours!

## Achieve life and career goals by building on what you know

The book begins by inviting you to reflect on your goals and priorities. With those aspirations in mind, examine your dissertation or thesis, as well as other writings from your doctoral studies or professional life. Identify strong pieces or sections of writing as *assets* that can be used on their own or developed further to fit a selected publication type.

Chapter 1 focuses on career and life goals, and what publications can help you to achieve. Chapter 2 gives some general guidance on ways to think about using your thesis or dissertation and other writings as a springboard for publications. At this point you will begin to outline your personal publication strategy you will develop through exercises in Chapters 3–11.

## Publication options and PIPS

Each of Chapters 3–10 focuses on different types of publications, explaining how you could use your identified assets to publish in a range of formats and on different platforms. Chapter 11 advises on how to find the right publisher for your projects and how to work with book and journal publishers. The book concludes with Chapter 12, which gives you the resources and exercises you need to complete and implement your publication strategy.

We recommend that you explore all of the options to decide what combination will best fit your goals. To illustrate ways potential publication types can fit with different post-doctoral career paths, we've created fictional characters Kris, Ella, and Nathan. We call them 'People in Progress' (PIPs). In the final Chapter 12, examples from real-life post-docs who used the typology explained in the book are included to inspire you. These fictional and real stories are intended to help you find your own path to publication.

## Contemplate options through reflection and discussion

Each chapter suggests questions for reflection that you can write about in your journal or discuss with others. As a researcher, you may have used a journal to record observations and memos throughout the process. Now, as you transition from being a doctoral student or post-doc to a published author, consider the value of the writer's journal. Keep track of the projects you are working on right now and ideas you might return to later.

You are not alone, since many students and graduates with PhD or professional doctorate degrees are similarly struggling to move forward. Use the exercises in this book with friends or colleagues who are trying to work through the publication process. Having a writing circle, or simply a person or two, that you can turn to for discussion and feedback, support, and accountability, will contribute to your success.

It can also be useful to consider in more detail the reasons why you want to write and publish. Writing and publishing is a way to effect change in the world. Like most ways of effecting change, it demands time and commitment. So why do you want to write and publish beyond helping your own career? Only you can figure out the answer – and taking that intellectual and emotional step will be worth the effort.

### *Take steps through exercises and complete a multi-year publication strategy*

The exercises included in each chapter allow you to examine writing and identify your assets, consider advantages and disadvantages, and plan key steps for each type of publication. When you reach Chapter 12, you will pull all of this work together and set the priorities and timelines needed for the publication strategy.

We recognise that some people may not choose to read all of the chapters and complete all of the exercises. However, you will need to complete the exercises in Chapters 1, 2, and 12 if you are to generate a publication strategy that can guide your work towards successful accomplishment of life and career goals. You are free to skip sections of this book that do not focus on publication types you plan to pursue at this time, though we would recommend at least skimming the text so that you're aware of all the options. As an added incentive to motivate careful attention to the important work of assessing and planning, the exercises in Chapters 1 and 2 should also help you with job applications. Ultimately, the more chapters you work with and the more exercises you complete, the richer and more useful your publication strategy will be.

The Routledge book website includes a workbook with all of the tables and exercises, organised by chapters. They are formatted as downloadable templates and worksheets you can customise and use throughout that process. These materials, plus additional resources and information about related events or activities with the authors, are also available on www.path2publishing.com.

## Advancing in your career in a competitive environment is challenging

Getting published can be a catalyst for increasing the impact of your research and developing a credible professional identity. Use the approaches described in this book to create a publication strategy that helps you to achieve the goals you had in mind when you launched your doctoral journey, or those that you developed en route.

# About the authors

**Janet Salmons, PhD.** I define my professional role as a 'free-range scholar'. This means my interests and curiosity are not constrained by an institutional affiliation. I work independently to do research, write, and coach, through my practice, Vision2Lead.com. I am the Methods Guru, and manager for SAGE Publications' research blog community Methodspace.com, where I write about using new methods, teaching research, and getting published.

I've focused my publication strategy on writing books. They include *Find the Theory in Your Research*, SAGE Publications (2019); *Getting Data Online*, SAGE Publications (2019); *Learning to Collaborate, Collaborating to Learn*, Stylus Publications (2019); *Doing Qualitative Research Online*, SAGE Publications (2016); and *Qualitative Online Interviews*, SAGE Publications (2015). I edited the *Cases in Online Interview Research*, SAGE Publications (2012). Forthcoming is *Collaborate to Succeed in Higher Education and Beyond*, with Narelle Lemon, Routledge (2020).

I learned a lot by serving as a doctoral faculty member, dissertation supervisor, and mentor to

PhD students from 1999 to 2017. I served on the Walden University College of Education faculty and the Capella University School of Business faculty. I was honoured with the Harold Abel Distinguished Faculty Award for 2011–2012 and the Steven Shank Recognition for Teaching in 2012, 2013, 2014, 2015, and 2016. I was pleased to be recognised with the Mike Keedy Award (2018) from the Textbook and Academic Authors Association, given in recognition of enduring service to authors.

A lifelong learner, I received a BS in adult and community education from Cornell University; an MA in cultural policy studies from Empire State College, State University of New York, and a PhD in interdisciplinary studies and educational leadership at the Union Institute & University. I live and work in the foothills of the Rocky Mountains in Boulder, Colorado.

**Helen Kara, PhD.** I have been an independent researcher since 1999 – that's over 20 years. I am not, and never have been, an academic, though I have learned to speak the language. I have taught research methods, including writing, to postgraduate students and staff in Europe, the Middle East, Asia, Australia, and Canada.

I have written various books on research methods and ethics, including *Creative Research Methods in the Social Sciences: A Practical*

*Guide*, Policy Press (2015); *Research and Evaluation for Busy Students and Practitioners*, 2nd ed., Policy Press (2017); and *Research Ethics in the Real World: Euro-Western and Indigenous Perspectives*, Policy Press (2018). In 2015 I was the first fully independent researcher to be conferred as a Fellow of the Academy of Social Sciences. My website is at https://helen-kara.com, and I blog most weeks at https://helenkara.com/blog.

# Acknowledgements

Janet would like to acknowledge Cole Keirsey for patient support of the writing life (and timely reminders to break once in a while). She appreciates her doctoral mentees for teaching important lessons about the challenges associated with moving from student to professional life, with particular thanks to Bessie Bowser, Donald Dunn, Joshua Fuerhrer, Kristine Pasley, and Lisa Toler for sharing their publication experiences. And special thanks to Helen Kara, for her valiant efforts to make this book a valuable resource for readers.

Helen would like to acknowledge her collaborators on this book and its series, Janet Salmons and Pat Thomson, who are a joy to work with and enrich her life.

We would both like to thank our beta readers, Brendon Fox, Amanda Heffernan, and Jon Rainford. Our text is stronger for their thoughtful feedback.

# 1 How can publications help build my career?

**After studying this chapter, you will be able to:**

- Identify your career goals and related publication expectations.
- Understand options for drawing from your thesis or dissertation.
- Evaluate ways to use your academic writings as the basis for publications.

## Overview

You know that getting work published is a positive career step. You build credibility by telling the world something about your research and your ability to develop new understanding of specific problems. At the same time, you share something about yourself, your experience, and observations. You demonstrate your ability to think critically and creatively. So, let's consider key questions: What kinds of books or articles, posts or media, will help you move forward? How can you use your thesis or dissertation and other academic writings as the basis for new publications? Answers to these questions are unique for each person and the goals they have for research impact as well as for their careers and personal lives. The premise of this book is that the types of publications you choose to produce, and the ways you shape the content of your writing, can be strategically selected in order to achieve life and career goals. Publications are beneficial to almost any career. If we hope to be hired by an educational institution, company, agency, or organisation, publications display skills and knowledge. If we are hoping to work independently, publications tell prospective clients something about our expertise.

**Reading this book and completing the exercises will enable you to:**

- *Assess* the publication potential of your thesis or dissertation and related writing.
- *Reflect* on scholarly or professional choices in light of your career goals, and *know* how to use publications to help you move closer to those goals.

- *Evaluate* steps to pursue traditional, online, or self-publishing: Understand the pros and cons of writing for blogs, academic journals, books and other formats.
- *Develop* a clearly defined two- to five-year publication strategy that aligns with your career goals.

## From research to publications

Rallis and Rossman (2012) point to three ways that findings are used:

- **Instrumental Use:** Research is designed to solve a problem or answer a question; findings are applied to provide solutions or recommendations.
- **Enlightenment Use:** Research findings contribute to a deeper understanding or insight that over time can inform decisions or actions.
- **Symbolic and Political Use:** Research findings provide new ways to represent a situation or phenomenon, serving symbolic purposes. Within this category, Rallis and Rossman highlight the importance of emancipatory uses of research to empower others toward social action, transforming social structures, practices, and relationships to improve society.

Research uses are political when used to shape legislative or policy developments (Rallis & Rossman, 2012).

These uses may overlap and you might employ two or all three in your work. However, this typology is a useful way to think about how we use research findings because each of these uses has implications for your publication strategy. Perhaps your research resulted in a solution to a problem in your field, so you will create publications spelling out guidelines and steps for implementing a new approach. However, your findings might not be clearly associated with a recognised problem, or the scale of the problem could be such that results from one study would be inadequate to address it. Instead, you might want to 'enlighten' other scholars in your field or the general public about how your study could contribute to a better understanding of the problem. If you studied a vulnerable population or a persistent social issue, your findings might have symbolic value by communicating a new vantage point. Depending on your field, you could be interested in public or institutional policy changes, and shape your writing accordingly.

The right choices for your publication strategy also depend on whether you want to start a new career, advance in your current career, change careers, launch independent freelance work, or start a business. While it might once have seemed obvious which types of publications will open which doors, today the lines are not so clearly drawn. For example, while those who aspire to take up full-time academic positions are typically expected to publish in peer-reviewed journals, they might also need to build readership by writing for a blog or the mainstream media. Freelancers or independent advisors who could be expected to write practical materials might also want to establish their scholarly credibility by publishing in academic journals. It is important to consider all of the options, and select which best fit your needs, interests, and priorities.

By writing professionally we establish a position within a field. That is, we create a personal identity or brand. For example, we might see ourselves as researchers or as people who interpret research results to practitioners in our field. We may feel our strengths are in the methods or theories used in our doctoral research. We might see ourselves as having particular expertise in the subject we studied or as having real-world experience that gives us special insight.

Through publications we pose and answer questions we believe are important, and present our analysis and interpretation of the evidence. Depending on the nature of the publication, the evidence we cite to support our position varies significantly. Evidence for a scholarly publication might be drawn from empirical research findings. When these are findings from our own studies, we convey a level of understanding of the research process. When these are findings from other studies, we communicate our knowledge of relevant literature. Evidence for a professional publication might be drawn from first-hand observations of the field and experiences in our communities or workplaces. We might try to bridge the scholarly and the professional milieus, and write in ways that encourage readers to apply lessons learned from the research.

We can write alone or in collaboration with others. While this book makes some reference to collaborating, there is not enough space for us to cover all the potential issues raised by collaborative writing. These are addressed in another volume in this series, *Collaborate to Succeed in Higher Education and Beyond* (Lemon and Salmons, 2020).

The choices about the kinds of writing you undertake can affect how others see you and your position within the field. Your professional identity can influence the sorts of professional opportunities you get such as invitations to consult, help someone write a grant proposal, or speak at conferences.

## Defining key terms

Lingo varies between different countries, so we want to clarify key terms used throughout the book. First, the terms *dissertation* and *thesis* mean different things in different parts of the world. For example, in the United States a Masters' student writes a thesis and a doctoral student writes a dissertation; in the United Kingdom it is the other way around. As this book is designed for late stage and post-doctoral students, we regard 'thesis' and 'dissertation' as synonymous, meaning the long piece of written work submitted for examination by a doctoral student. Most of the time we will use both terms, though at times we will use one to mean both. We use the term *PhD* to refer to doctoral study generally, including both traditional scholarly programmes as well as professional doctorates such as the education doctorate (EdD) or doctorate in business administration (DBA).

Masters-level curricula also vary greatly. Some Masters' students complete original research and write a scholarly paper while other students complete a more practical final project sometimes called a *capstone*. While the book focuses on doctoral-level students and postgraduates, we would also encourage readers who completed significant research and writing at the

Masters level to evaluate how that work could be used as the foundation for publications.

Another variation in terminology: In the United Kingdom, *school* is where you go for your compulsory education. If you wish to continue on to higher education after school, you go to *college* or *university* (although UK universities tend to have a department called a *graduate school*, just to confuse things!). In the United States, people talk about 'going to school' when they mean education at any level.

## Thinking about the thesis or dissertation

'A dissertation [or thesis] is an extended scholarly document written to describe an original and independent research study conducted by a doctoral student after completing all coursework and in partial fulfilment of the requirements for the doctor of education or doctor of philosophy degree' (Frey, 2018, p. 535). While there are a variety of styles, for the purpose of this book, we focus on the most typical five- or six-chapter format. This kind of thesis or dissertation will include chapters as follows:

1 **Chapter 1** introduces the study, including the description of the research problem,

questions or hypotheses, an overview of the methodology and methods, and the theoretical framework.

2 **Chapter 2** offers scholarly foundations for the research through a literature review, and identifies gaps in existing knowledge that the study intends to fill.
3 **Chapter 3** provides an in-depth description of the research design, including methodology, methods for data collection and analysis, strategies for sampling and recruiting human participants or other sources, and discussion of ethical issues.
4 **Chapter 4** reports on the data and results of the data analysis.
5 **Chapter 5** summarises the study and situates the results in the empirical literature. The thesis or dissertation concludes with implications of the study for future scholarship and practice.

Depending on the institution or the nature of the study, these basic chapters can be expanded. A six-chapter dissertation or thesis could include two chapters that report on data. For example, a mixed-methods doctorate might merit Chapter 4 to discuss qualitative analysis and Chapter 5 to discuss quantitative analysis. Or, a six-chapter dissertation or thesis could include two chapters that discuss the findings. For example, Chapter 5 could be dedicated to contextualising the findings in the literature, and Chapter 6 could focus on implications and significance of the findings. The key point here is that each part of the document contains material you can draw on and develop further for publications.

### *Find your assets*

You can use your writings from any and all of your chapters in a variety of ways for academic or professional publications. Throughout your Masters and doctoral studies you have completed additional writing that you can also mine for content. These could include annotated bibliographies or literature reviews, conference presentations or posters, substantial research papers, presentations or slide sets, and government documents or reports. You might have created visuals, tables, diagrams, or media. One of the first exercises we provide at the end of this chapter is a thorough audit of your existing work so you have a comprehensive picture of the strengths and limitations of the writing you have completed. We call these pieces of writing and visuals your *assets.* You will learn to think of ways your identified assets can serve as the foundation for the next stage of your life and career.

## Moving from student to professional writing

Writing you complete as a student serves a specific purpose: to demonstrate your competence in the context of your programme. In Masters' level courses you are expected to show that you are able to read, interpret, and discuss empirical research as presented in scholarly articles. Most Masters-level writing is aimed at a limited audience, generally fellow students for team assignments or peer interactions and your faculty members. It is unusual to have an audience beyond the classroom, which means all your readers have a common understanding of the topic at hand and a similar background in the literature.

When you move into doctoral research, the expectations increase and the audience expands. You are expected to show that you comprehend and can synthesise concepts from seminal and emerging research in your field. You must apply what you've learned by coming up with an original research question, and design a study that will address that question. The immediate readership may include not only your supervisor(s) and committee members, but also other reviewers within the university. You may have published some articles based on your research design, literature review, or even your findings.

If you are a recent doctoral graduate, your discussions of the literature and ideas about issues in your field as presented in course papers may still be relevant. If you have been out of education for a while, the bedrock scholarship in your field still provides a foundation for current thinking, but you might need to enhance earlier writings with more current citations.

It is beyond the scope of this book to talk about how to improve your writing style. We offer some resources to help with this in our recommended reading and on the book's companion website. Also, it can be hard to assess the quality of your own writing, but here we can offer some criteria to help. These were developed by Sarah J. Tracy of Arizona State University to assess the quality of qualitative research and they also work well to assess the quality of writing. Tracy suggests eight criteria (Tracy, 2010):

- Worthy topic
- Rich rigour
- Meaningful coherence
- Resonance
- Sincerity
- Credibility
- Significant contribution
- Ethics

Tracy (2010) suggests that the topic on which you write should be worthy, that is, valuable, relevant, and timely. Use this criterion as a springboard for related questions: For whom is the topic worthy and valuable? If the topic is not relevant and timely, how might you update it?

In writing, rigour means taking care over all the elements of the writer's craft from word choice to overall structure, with the aim of using the options most likely to meet your readers' needs. Writing that has meaningful coherence tells an intelligible story, weaving together diverse elements such as literature, theory, data, and analysis in a comprehensible way. Use these criteria as a springboard for related questions: How would you characterise your writing style? Do you need to refine it to better reach your target audience? Will you need to do rewriting or reorganising because the original thesis or dissertation lacks coherence or because you have added updated literature?

Resonance means your writing has a positive emotional impact on readers by offering them information, help, or inspiration. Sincerity means honesty and credibility means trustworthiness. Use these criteria as a springboard for related questions: Will your writing resonate with your target audience? Is the academic style of writing, typical for a thesis or dissertation, too stiff or formal for your new context?

Do you need to add images or stories to convey sincerity?

In terms of writing quality, the term *significant contribution* can seem like a stretch goal for a new writer. At its most basic, it means the content of your writing is original, that is, it hasn't been published in the same form elsewhere. A significant work contributes to others' scholarship, provides new insights into problems, or offers new approaches, practices, or policies. A significant contribution exemplifies ethical practices in research and writing. Ethical writing is truthful, clear, and cites the work of others appropriately; this is discussed in more detail in Chapter 2.

You can use Tracy's eight criteria as benchmarks to assess the quality of your writing.

## Steps for publishing start with clear goals

Steps for publishing in advantageous ways vary greatly depending on your field, discipline, and your life and career goals. Before you can create a proposal or draft you need to know what you want to do and why. Start by reflecting on what you've accomplished in your Masters, doctoral, and professional work,

**Initiate Your Publication Strategy**

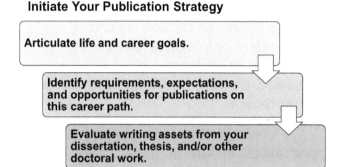

*Figure 1.1* Initiate your publication strategy.

and what you want to achieve (Figure 1.1). What expertise and knowledge do you have, and what do you want to present to the public broadly and your field in particular? On what do you base your credibility: your research, your writing, teaching, or what you've learned from experience in life and professional activities?

Typical careers for doctoral graduates include (but are not limited to) the following:

- Tenured academic
- Nontenured instructor
- Academic administrator
- Leader of a government agency, nongovernmental, or non-profit organisation

- Professional or social researcher in an institute, corporate, or community setting
- Independent or freelance researcher
- Consultant or advisor
- Business-person, entrepreneur, or social entrepreneur
- Writer

Opportunities in any of these career options would be boosted with publications that demonstrate your knowledge and expertise. The options for research dissemination continue to expand, while at the same time, competition is stiff for the attention of readers who suffer from information overload. To reach readers and make an impact, it is unlikely that focusing on one type of publication will be adequate. While it might seem obvious that those who are interested in a tenured position would focus primarily on scholarly articles in peer-reviewed academic journals, they also may need to connect with readers through blog posts or the mainstream media. While it might seem obvious that freelance researchers or consultants could benefit from publishing a guidebook or manual, they may also need to build credibility by publishing scholarly research in academic journals. In the process of working through this book, you will develop a strategy and create a plan that will most likely include more than one type of publication.

## Matching goals with publication options: people in progress

What does this process of reflecting on life and career, goal-setting, and prioritising options look like in real life? Throughout the book we will point to different concerns and choices with stories that exemplify common situations faced by people who want to use publications to build or advance careers. We are calling these fictional characters, Kris, Ella, and Nathan, 'People in Progress' or PIPs. They are pursuing three different life and career paths: Kris wants a traditional academic position, Ella wants to craft an independent research niche, and Nathan wants to conduct research in an agency or institute. Kris and Nathan have both received their doctoral qualifications while Ella is close to the end of her doctoral study. Whether or not your aspirations and situation are similar to one of the PIPs, their examples will be helpful as you discover that there is not a single publication strategy that fits all.

To further reinforce the point that we each need to personalise our respective publication strategies, in Chapter 12 you will hear from a group of Janet's former doctoral students. They were exposed to the ideas presented in this book and have tried to put them into action. We hope that the fictional PIPs and the real doctoral graduates will inspire you to think creatively about the directions you want to take.

Let's meet the PIPs we'll follow as they take steps towards getting published:

**Kris** hopes to be a professional academic. At 35, with eight years as a civil servant before doing a Masters' degree and falling in love with academia, Kris has an understanding of diverse social policy needs. Doctoral research on the difference between the way taxes are spent in the United States and the way Caucasian, African-American, American Indian, and Hispanic Americans would choose to spend their taxes generated both theoretical and practical findings, including a small but significant advance in quantitative methodology. Kris has secured a one-year post-doctoral contract in an academic department that is well regarded in their field.

Publications are essential to finding a tenured position. Kris' initial goals include publishing an article in a respected journal, and Kris assumes that to continue advancement into a full-time faculty position, they will need to continue to publish in top journals. They would also like to network with other new and established researchers by contributing to a well-read blog managed by a professional association in the policy field.

**Ella** is committed to a career as an independent scholar. She is 53 and used to be a schoolteacher. For her doctoral research, she is doing a phenomenological study of the school experiences of lesbian, gay, and bisexual teenagers (she would happily have included people who identified as trans, intersex, asexual, etc. but all her participants identified as either lesbian, gay, or bisexual). Ella became an independent researcher because she felt she could do more to support LGBTQIA+ schoolchildren from outside the system than she could within it. She likes her independence, though she needs to maintain and develop links with schools and higher education to ensure that she gets enough work.

While scholarly publications are helpful for anyone who wants to work with education professionals, particularly those in higher education where publication in respected arenas is highly valued, Ella also wants to produce one or more publications to help people like her participants and their families. However, as an independent researcher, she has to balance her publication goals with her need to earn a living.

**Nathan** intends to build a career as a psychiatric nurse researcher for a research institute or government agency. He is 31 and has been a psychiatric nurse all his working life. For his doctoral research, he investigated the ways in which people can maintain good mental health while a close family member is on overseas military service. Nathan is passionate about mental health and about research. He wants more people to understand the importance of good mental health and how to achieve it, and he wants his colleagues to apply new research findings in their profession. He enjoyed his doctoral studies but has no wish to pursue an academic career.

Publications will help Nathan to communicate his passion and make a difference in people's lives. His goal is to publish as widely and as accessibly as possible.

Perhaps you identify with Kris, Ella, or Nathan, or perhaps you don't identify with any of them. Either way, you will find useful tips in later chapters by reading descriptions of how they use their theses, dissertations, and other academic work to plan publication strategies that will help them achieve personal and professional goals.

## Chapter summary

Publications can help you to make the transition between academic and professional life. However, in an era when publishing is rapidly changing, it is not always obvious what types of posts, articles, or books will be most beneficial. By reflecting on and clarifying your goals, and understanding the patterns of knowledge exchange in your field, you can use your time wisely. By looking critically at not only your final thesis or dissertation, but also other academic and professional writings you have completed, you can identify insights, areas of expertise, and supporting evidence you can use as the basis for formal or informal publications.

## Get started!

You have hopes for the first – or the next – stage of your career. It is time to articulate your goals and examine specific ways publications can help you fulfil your dreams. You have knowledge and insights that can benefit others, but your contribution will never be made fully if your files stay buried on your hard drive. Pull them out, take a fresh look, and think critically and creatively about how your hard work can bear fruit.

All of your research and writing provides you with invaluable material you can build on in a variety of ways. The exercises for this chapter invite you to audit your existing collection of writing, to imagine and contemplate your career goals, and to think through the expectations or requirements for publications associated with them. What types of writing and what kinds of publications will you need to achieve your goals – and make the contributions that will advance your career? What knowledge, strengths, and personal identity will help you move forward? Continue to refine and develop your goals as you move through the chapters and exercises in this book.

In the process of working through this book, you will develop a publication strategy. In the exercises for this chapter, you will start by reflecting on what you've accomplished in your doctoral and professional work, and what you want to achieve. What expertise and knowledge do you have, and what do you want to present to the public broadly and your field in particular? On what do you base your credibility: your research, your writing, teaching, or other professional activities? Your answers will serve as the basis for developing your publication strategy. You will add elements to the strategy in Chapters 2–11. In Chapter 12 you will prioritise and sequence the steps you want to take, and think about ways to stay on track.

Visit the Routledge book website and www. path2publishing.com for templates and worksheets you can use to complete this chapter's exercises. You will find that exercises and questions are available in a usable workbook that is organised by chapter for easy reference.

## Exercises and questions for reflection: identify and align goals, assets, and plans

These exercises ask you to reflect on your career goals and align them with your current assets. Reflection questions encourage you to think first, and then work through the exercises (Figure 1.2). For our purposes, the term *asset* is used to refer to any piece of relevant work, whether it is written, visual, or in other media.

Feel free to add or rename columns and to customise the worksheets to fit your own circumstances. You may find you need to move back and forth between them as you identify and align goals, assets, and plans.

### Reflection

Why do you want to write? This is the fundamental question. Use this chapter's exercises and questions to think through the reasons and aspirations that motivate you to write. Discuss your thoughts and feelings with other writers and/or record your observations in a writer's journal.

Reflect on your career direction(s). Are you trying to advance in your current career or change careers? Are you looking for a position in an academic institution, a company, or a charity, nonprofit, government agency, or nongovernmental organisation? Do you intend to create your own business, such as a consulting practice? Do you want to be entirely self-employed, or to combine selected options with freelance work that complements a salaried position? Make a note of the route you want to take.

## Evaluate Assets:
## Review Writings, Reflect on Experiences!

### Academic
- Thesis/Dissertation
- Lit reviews, papers
- Presentations

### Professional
- Projects
- Reports

### Personal
- Travel
- Unique professional or cultural experiences

*Figure 1.2* Evaluate your assets.

What are the personal dimensions, perspectives, or priorities as you look at your career goals? How might your hopes to travel, start a family, or other personal desires factor into your career plans? How might you integrate activities you've done as side hustles or hobbies into your work and writing life?

### Exercises

*Exercise 1: articulate life
and career goals*

Make some notes about what you know at this point in regard to expectations for types of publications associated with your preferred career option (Table 1.1).

If you aren't yet sure what might be expected, begin with some research. First, look for the highly successful thought leaders in your field. Look at their publications, as well as social media, blogs, or videos. Check out their institutional affiliations and other professional and volunteer activities. What can you learn about the kinds of efforts valued in your field? Next, identify your questions and ask for more information from academics, professionals, or leading members of professional associations in your field.

*Table 1.1* Career goals and publishing

| Career Goal | Writing and Publishing Needed to Achieve Each Goal |
|---|---|
| Tenure-track academic | Peer-reviewed articles in journals respected by your field |
| Freelance blogger on research methods | Books and/or articles about research methods |
| | Portfolio of short essays and blog posts |
| Consultant | Publications that showcase your abilities and showcase usable models and strategies |

## Exercise 2: identify and analyse the assets that will help you to meet your goals

What types of writing and what kinds of publications will you need to achieve your goals – and make a contribution? What knowledge, strengths, and personal identity will help you move forward? What prior research can you draw on as content for new writing? What previous writings could be updated and fleshed out for publication? This is the first stage of the publication strategy development process: to clarify what is needed to meet personal and career goals, and to dissect and evaluate your existing work.

This exercise and the next will help you to collect, organise, and evaluate all of the materials you have. Of course, you have your thesis or dissertation, but you may also have other substantial assets that you haven't considered in the same light. To get started, dig through your filing system and pull out significant papers you wrote for relevant courses, as well as any previous articles (published or unpublished), essays or assignments, research reports, conference abstracts/posters/papers, relevant social media posts, and so on. Consider any professional work, such as reports, articles, presentations, or webinars. List these under the headings below, deleting any headings that are not applicable for you.

## Exercise 3: identify your assets

Begin by creating a list of relevant academic writings from coursework, your doctoral programme, postgraduate, and/or professional efforts. They might include:

- Thesis or dissertation
- Course papers

- Articles
- Essays or assignments
- Research reports
- Professional reports
- Conference abstracts, posters, papers
- Funding applications
- Presentations, media, and/or webinars
- Relevant blog or other social media posts
- Other

## *Exercise 4: evaluate your assets*

Next, carefully evaluate each of these intellectual assets. What ingredients do you have, and what will you need to develop, in order to cook up the career you want? Create as many tables as you need to describe and analyse your work. You might want to use one table for an entire dissertation or thesis, course paper or essay, or create tables for specific sections of the dissertation/thesis, such as the literature review. Try to identify at least three assets you can use; it's fine if you have more. Use the worksheet from the website and add criteria and columns to customise the table to fit your needs.

Here (Table 1.2) is an example we have completed to show you how this works:

Table 1.2 Asset example

| Dissertation/Thesis, Title: xxx has: | Section/Pages | Description | Author's Notes |
|---|---|---|---|
| Strong and clear chapters or sections? | Chapter 1: Introduction to the problem | Problem is described very clearly | I should add a couple of specific examples from the field, in addition to current literature. |
| Points are relevant to current issues in the field or research and/or practice? | Chapter 1: Background of the problem | Problem is still relevant | |
| Literature is up-to-date? (Less than 3–5 years old depending on field of study.) | Chapter 2: Literature review | Most recent literature is 2014 – need to add new sources | With minimal updating, I could use the introduction to the problem as the basis for a short article or even a blog post. |

*(Continued)*

Table 1.2 Asset example (Continued)

| Dissertation/Thesis, Title: xxx has: | Section/Pages | Description | Author's Notes |
|---|---|---|---|
| Data are current (less than 3 years old). | | N/A in this section I don't discuss my data | |
| Raised new questions for future research, the field of study, society at large? | Chapter 5: Implications of the study | New questions point to aspects of the problem that are still unresolved | I could add specific recommendations to be more practical. |
| Suggested recommendations for practice? | Not included in the dissertation | | |
| Other | Paper from xxx course: 'title', pages 5–8. | I discussed the problem from a societal perspective in this paper. | Question: Where could/ should I publish it? |

## References

Frey, B. B. (2018). Dissertations. *The SAGE encyclopedia of educational research, measurement, and evaluation.* Thousand Oaks, CA: SAGE Publications, Inc.

Lemon, N., & Salmons, J. (2020) *Collaborate to Succeed in Higher Education and Beyond.* Abingdon, Routledge.

Rallis, S. F., & Rossman, G. B. (2012). *The research journey: Introduction to inquiry.* New York, NY: Guilford Press.

Tracy, S. J. (2010). Qualitative quality: Eight "big-tent" criteria for excellent qualitative research. *Qualitative Inquiry, 16*(10), 837–851. doi: 10.1177/1077800410383121.

# 2 How can I create new publications based on academic writings?

**After studying this chapter, you will be able to:**

- Identify ways to use your identified assets as the basis for publications.
- Align your career and publication goals with your existing assets.
- Start developing a publication strategy.

## Overview

Getting your work published is a multistage process that starts with a purposeful plan. Careful reflection and systematic analysis are critical in order to make the best use of your research findings and the new knowledge you acquired in your doctoral studies. We call this process *creating a publication strategy.* A publication strategy should include carefully defined goals, a purposeful timeline, and actionable steps for proposing and writing the kinds of pieces – large or small – that produce impact, allow others to access what we've learned, and propel our careers forward.

> A *publication strategy* is a thoughtful plan of action for aligning life and career goals with publication options.

Creating a publication strategy helps you to think through what you want to publish, how, when, and in what formats. We encourage you to outline a two- to five-year strategy that encompasses both formal and informal types of publication. Chapters 3–11 of this book will help you to choose between different publishing options. Completing the exercises at the end of each chapter in this book will enable you to create your publication strategy. Once you have the strategy, you can focus on writing projects in a sys- tematic way. Of course, that is still a big chunk of work, but it's much easier to focus on writing when you have already done the thinking about what, how, when, etc.

Attention to clear, ethical writing is essential as you transition from student work to publica- tion. Plagiarism, self-plagiarism, or unintentional errors in citing sources can be problematic when you are in education but can destroy your career when included in manuscript submissions. Most contemporary publishers use the same plagia- rism detection software now common in colleges

and universities – and you do not want to jeopardise your efforts to be an academic writer with
problems of this kind.

Once you have studied all the chapters of this
book and completed the exercises, you will not
only have a strategy covering a suitable number
of years but also have set interim milestones to
help you stay on track. The exercises in Chapter 1
invited you to imagine and contemplate your career
goals and to think through the expectations or
requirements associated with them. You audited
your collection of writings and other materials to
identify your assets. What interesting observations emerged from your analysis that you could
build on, adapt, or develop further? Let's dig a little
deeper and begin to consider publication options
and ways to ensure ethical, trustworthy writing.

## Five ways to use your identified assets

Completion of your doctoral work is a triumph
of effort and persistence wherein besides academic writings for papers and projects, you finished your thesis or dissertation. Roberts and
Hyatt observed that:

> Completing a dissertation represents the
> pinnacle of academic achievement. It
> requires high-level skills of discernment

> and critical analysis, proficiency in at least one research method, and the ability to communicate the results of that research in a clear, coherent, and concise manner. (Roberts & Hyatt, 2018, p. 20)

The thesis or dissertation is written to satisfy the requirements of your academic institution, supervisor(s), and committee. It is written in a specific format and style as needed to demonstrate your mastery of the stages of research design and conduct, and your ability to write about the significance of the completed study. (See Chapter 1 for more about typical thesis/dissertation formats.) This is generally not the same kind of writing style and form needed for postgraduate publications. In most cases, some degree of rewriting and reformatting will be needed. That said, all of your identified assets, including the chapters of your thesis or dissertation, are a treasure trove you can draw on in five distinct ways.

When co-author Janet Salmons was trying to encourage her doctoral students to consider the whole dissertation as content for publications, not simply the results of the study, she reflected on her own experiences. One of her first forays was a set of definitions included in a technical encyclopaedia – definitions drawn from Chapter 1 of her dissertation. By adding specialised literature and conducting a set of expert interviews, she was able to expand on and adapt her findings for a chapter in an edited collection (Salmons, 2007). While discussing plans to write about her findings, she discovered that other academics were intrigued by the methods used, so she expanded on Chapters 3 and 4 of the dissertation for what became the first in a series of books about online research (Salmons, 2010). To communicate the range of options, she created the typology shown in Figure 2.1. We will refer to this typology throughout the book and use it as a framework for discussing a wide range of publication options.

You could *extract* an element that could stand alone. Additional writing might be needed to introduce the piece, or to update sources if your research is a few years old.

---

**Extract**

- Determine which part(s) of your academic writing could stand alone.
- Plan how you will develop the selected extract(s) into a publishable blog post, essay, article, or chapter.

---

As you look through your identified assets, look for particularly strong sections that you can use as is or build upon for a new piece of writing.

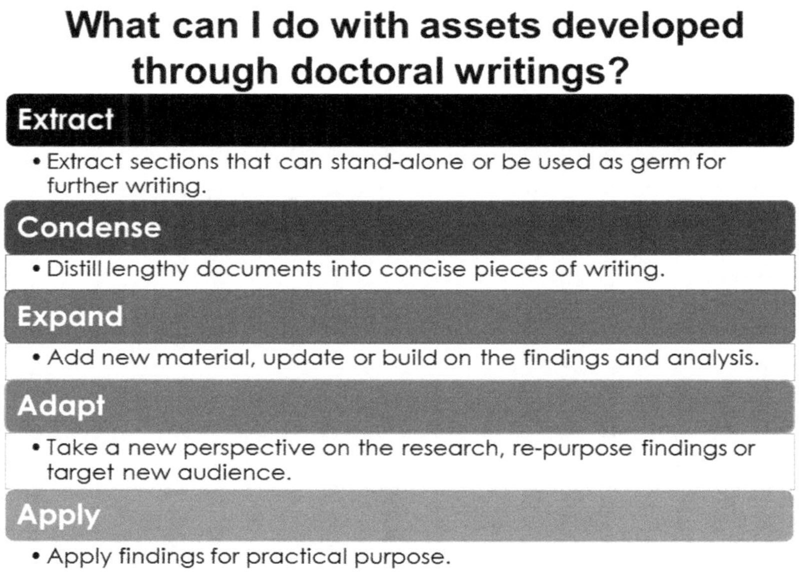

# What can I do with assets developed through doctoral writings?

**Extract**
- Extract sections that can stand-alone or be used as germ for further writing.

**Condense**
- Distill lengthy documents into concise pieces of writing.

**Expand**
- Add new material, update or build on the findings and analysis.

**Adapt**
- Take a new perspective on the research, re-purpose findings or target new audience.

**Apply**
- Apply findings for practical purpose.

*Figure 2.1* Doctoral publication typology.

For example, you could extract a statement about the problem you investigated and write a blog post about it. You could extract your literature review and publish it as a journal article. You could extract a section about your methodology and publish a chapter in a book about research methods. You could extract sections to help you write about the data or your analytic strategy. You could extract sections to help you write about your findings and their implications.

You could condense the entire thesis, dissertation, or other significant academic paper into an article.

---

### Condense

- Look at the main ideas in your identified assets. Which idea lends itself to an article or chapter?
- Plan how you will select and edit this large piece of work into one that fits length constraints.

---

The structure of a five-chapter doctoral thesis or dissertation corresponds roughly to the structure of a standard scholarly journal article that reports on research. Both include an introduction that states the problem and provides an overview of the study, a review of foundational literature, a description of the research design, an explanation of the data and analysis, and a discussion of the findings and their implications. However, pages in the thesis or dissertation must be condensed into paragraphs or sentences in a much-shorter research article. Similarly, you could condense other significant academic papers or reports into shorter pieces of writing for formal academic articles or more informal posts or essays.

Alternatively, you could take the opposite approach and expand on sections of the assets you identified.

---

**Expand**

- Is your research up-to-date? If not, what new findings could enhance your work?
- What exemplars or cases, tips, or suggestions could you add to make the publication more robust?
- Plan how you will find or create new content to build on existing work.

---

Again, the institutionally mandated requirements for doctoral writing mean there are some topics that must be covered. There are other topics that you don't have time or room to develop, or that you are discouraged from including.

After graduation, you can add those elements to expand the reach and scope of your publication.

The identified assets may have been sitting in files while you got your career started or took care of personal aspects of life that were set aside while you were studying. If you are in this situation, you can expand on the original work to bring it up to date. This might entail citing newly published literature, analysing your data using a different method, doing some follow-up research so you can include recent data, or adding new insights into your analysis.

Another way that writers expand on their work is by collaborating with other researchers. If you join forces with one or more people to write books or articles, it is likely that you will expand your thinking to encompass their ideas and contributions.

Or, you can adapt your research to meet new needs or reach new audiences.

---

**Adapt**

- Review your research by taking a new look at the study findings and implications.
- Consider different ways to present your research.
- Think about new audiences or uses for your findings.

As noted, the thesis or dissertation must be very focused and specific to meet the doctoral completion requirements of your institution. There could be other ways to think about your work. For example, you might have studied personal hygiene from an educational stand-point but could adapt your writings to look at personal hygiene from a multidisciplinary standpoint.

Initial readers for the thesis or dissertation are faculty members and scholars who use a particular vocabulary and way of communicating. Similar principles apply for other assets such as conference papers or posters. To reach readers outside of academia, you might need to adapt your writing to a less formal style, or provide additional explanations and definitions of key terms. Non-academic readers will want to know more about what the study accomplished and what the findings mean, and less about theory and methodology.

Identified assets are typically written documents, perhaps with a limited number of tables or figures. Could you adapt your work to share it in other ways, including recording media or podcasts? Could you adapt your work to a visual form for a graphic book?

You can focus on applying what you've learned in practical ways.

**Apply**

- What impact can your discoveries make in your field, your community, or the world?
- Who could use your findings? How? What would they need to know?
- What practical guidelines or instructions would help someone to apply what you learned in your study?

This way of looking at the assets accumulated through your programme of study means asking difficult questions: *so what?* And, *who cares?* Was your study an intellectual exercise only or an exploration of real-world problems? Can your findings benefit those who are working in your field, making policies, trying to improve the environment or society, teaching students, or raising families?

The first chapter of your thesis or dissertation typically includes a description of the intended significance of the study; the final chapters typically include a section focused on the implications of the study. Can you use these parts as the foundation for writings that could enable others to use your recommendations in practical ways?

## Writing to reach your target audience

When thinking about the appropriate publication type and the impact you aim to achieve, consider your readers. Moving outside of your doctoral programme means you move outside a culture with a common understanding of terms and acronyms. Writing well and clearly is essential so people from other professional or global cultures can read and understand your work as easily as possible. Your readers could include:

- Academics, scholars, and/or students from your own discipline or related fields. This audience will be most interested in the ways you designed and conducted the study, the scholarly and theoretical foundations for the study, the data analysis, and results.
- Public or private policy-makers. This audience will be most interested in ways your findings could inform decision-making about new policies or regulations (Rallis & Rossman, 2012).
- Professionals or practitioners. Rallis and Rossman (2012) refer to this audience as 'those who contribute to getting the work done' (p. 162). This audience has real-world experience with the topics you studied. To get the work done, they probably have a college

degree or similar training, so they have prior knowledge in the field. This varied audience is looking for new approaches and practical tools they can use.

- The general public. These readers may have no prior knowledge of the field but are interested in learning (Luey, 2010). Memoir, histories, social commentary, and creative nonfiction are types of writing popular with non-academic audiences.

### Adjust your writing style

Each publication type discussed in this book asks you to write in a particular way. There is still a misperception in some quarters that academic writing necessarily requires lots of long words in long sentences – the more complicated, the better because that's what clever people do, right? Wrong. There is a famous quote from Albert Einstein: 'If you can't explain it simply, you don't understand it well enough'. Making your writing short, simple, and clear is actually more difficult than writing complicated convoluted prose. Another quote, attributed to Mark Twain, is: 'I didn't have time to write a short letter, so I wrote a long one instead'. The smart writer would take the time to write the short letter and to explain its contents simply.

Using the principles of plain English can help. Briefly, some relevant principles are:

- Be concise
- Use short words where possible
- Aim for an average of 15–20 words per sentence, and certainly no more than 25
- Vary sentence length
- Avoid jargon
- Always give a definition of any technical term you need to use
- Use active verbs, not passive verbs (e.g., 'she collected the data' rather than 'the data was collected [by her]')
- Write as if you are speaking to your reader
- Write in a way that is helpful and polite (Kara, 2017, pp. 193–194)

## Ethical writing

Many doctoral students are taught that ethics is only, or primarily, relevant when you are gathering data. However, ethics underpins all aspects of the research process, and writing is no exception. You are aware of the prohibition against plagiarism and of encouragement to cite, paraphrase, and attribute others' work correctly. Proper attribution is absolutely critical in any manuscript you submit for publication.

In your writing, you will be representing the work of other scholars. You have an ethical responsibility to read this work carefully, understand it fully, and represent it accurately. Otherwise you may misrepresent someone else's views, which could harm their professional reputation, or yours, or both (Luey, 2010, p. 88).

You will also be representing the input of your participants. If you use direct quotes, should you use a pseudonym? What if your participants object to you using a pseudonym? If you are going to use a pseudonym, how will you do this ethically, given that many names have connotations of gender and ethnicity? Should you ask participants to choose their own pseudonyms (Kara, 2015, p. 123)? These are some of the ethical questions to consider when deciding on the use of direct quotes in your work. And whether or not you use direct quotes, you should do justice to your participants by ensuring that you represent your data accurately and interpret your findings fairly.

There may also be ethical considerations that matter to you in respect of where you choose to publish. Some scholars are passionately committed to open access on ethical grounds, and publish everything they can in open access journals or via mainstream or social media or self-publishing. Others have their reasons for choosing industrial publishers but, for ethical reasons, select those that are independent and/ or non-profit rather than those that pay dividends to shareholders. Some scholars are also activists who want their work to contribute to social change or improvements. They have interests such as gender or racial equality, climate change, and economic fairness that extend beyond the boundaries of their academic disciplines. They want their writings to have an impact in the real world, including developing countries and the global south. For such writers it is important to consider the reach and accessibility of their published writings. These considerations should be part of your publication strategy planning process.

## PIPs

For **Kris** publications are essential to finding a tenured position. After carefully reviewing their dissertation, Kris decided on an initial set of options that they will work on while developing a comprehensive publication strategy.

First, their research problem statement was clear and succinct and pointed to an important need in their field. They will *extract* and *expand* it to include an introduction and summary. Kris will submit this piece to a newsletter published by their professional association. They will also *adapt* this description of an important research need into a presentation for the next association conference. Kris hopes these steps will help make connections with others who share this research interest, and open doors for a faculty position.

The second option entails *condensing* the literature review into a publishable journal article. Kris noted that positive feedback from the dissertation committee highlighted the unique slant they took by examining international, interdisciplinary literature. However, the literature review chapter of their dissertation was too long and detailed for an article. Kris decided to focus on tax-related literature for submission to a business journal. After condensing the chapter, they also needed to *expand* it to include an introduction and summary.

**Ella** thinks publishing from her doctoral study will help establish her reputation as an independent scholar. Her main goal is to *adapt* and *apply* findings as the basis for a series of books for LGBTQIA+ children and their parents. However, she has to finish her thesis first, but as soon as she has submitted it, she intends to write an article about her study for an education-oriented journal.

Also, Ella realises that to establish herself as an independent scholar, she needs to demonstrate her capability as someone who understands current research trends, and can design and conduct research. She plans to *extract* the parts of her thesis that centre on her research design.

She will *expand* this material into a substantive book chapter for an edited collection on how to conduct research with marginalised groups.

**Nathan** thinks publishing from his doctorate can help towards achieving his ambitions. His goals include *adapting* his research for practitioners and *applying* his findings in a guidebook for members of the psychiatric and counselling professions. He would like to find new ways to share what he has learned with patients and professionals. He would like to *expand* on the implications of his findings set out in his dissertation and *adapt* them in a podcast series aimed at his target audience.

## Chapter summary

This chapter invites you to think about how to align different parts of the work you started in Chapter 1. You are encouraged to evaluate which elements can stand alone (extract), which long sections can be summarised or shortened (condense), where you can add to or update existing pieces of writing (expand), how you can use research findings or writings in new ways (adapt), and/or how you can make a difference by showing how research findings can be used (apply).

By bringing together your reflection on professional and personal goals, and your analysis of the assets you already have in your existing writing, you are taking important steps toward developing a workable publication strategy. In the process, we encourage you to commit to publishing clear, ethical writing that will make a contribution to your field.

## Exercises and questions for reflection: plan to use your existing body of work

Use the reflection questions and exercises in this chapter to start thinking strategically about what you can do with the assets you identified in the Chapter 1 exercises.

Visit the book website or www.path2publishing.com to download the workbook that includes templates and worksheets you can adapt for these exercises.

### *Reflection*

Writing is a time commitment even when we have significant assets to build upon. What do the options for publication mean in the context of your personal priorities? What are you willing to sacrifice to make the time needed to complete writing projects at the level of quality needed for academic or professional publishing?

## Exercises

### Exercise 1: align assets and goals

In Chapter 1 you identified your goals and evaluated your academic writings and related work. Begin making plans by identifying assets you currently can direct towards meeting the goals. Create a column for notes about each goal, including next steps.

Table 2.1 shows an example of assets and goals.

### Exercise 2: academic work to publication

Use the five options explained in this chapter to think through potential directions you can take.

What can I do with my thesis or dissertation? Try to align publication options with the goals and assets you outlined in the Chapter 1 exercises. Describe whether you will extract, condense, expand, adapt, or apply some elements from your dissertation. Table 2.2 shows an example of goals and strategies.

How and when will you do the work outlined in the to-do list? Which steps can be done quickly, such as finding out what informal publishing options are available to members of professional associations you have joined? Other steps will take more time, such as finding, reading,

Table 2.1 Example of assets and goals

| Goal | Current Status | Plans or Steps | Assets | Notes |
|------|----------------|----------------|--------|-------|
| Become an independent researcher and adviser who conducts programme evaluations. As an independent researcher and consultant, I would be expected to create reports and, possibly, articles for professional newsletters or journals. | I conducted evaluations and wrote reports for a student research project; studied the evaluation techniques in my dissertation research. | 1 Write an article about programme evaluation methods in community organisations. 2 Post about the article on a blog aimed at agency staff, link to social media. 3 Use this article and related posts to build credibility, expertise in programme evaluation. | Reviewed the literature and wrote two academic papers on evaluative methods for a PhD level class. | The literature I reviewed is now three years out of date, so the first step is to update the literature. I need to publish an article that would build credibility in my expertise. Ideally an article should be published in a journal respected by both scholars and agency leaders. |

Table 2.2 Example of goals and strategies

| | | Dissertation/Thesis Title | |
|---|---|---|---|
| Option | Description | Related Goals | To-Do List |
| 1. Extract | Use introduction to the problem as the basis for a blog post. | Build credibility for my expertise | See whether my professional association has a blog where my post might gain attention from others in my field. I might ask whether anyone would like to collaborate on an article. |
| 2. Condense | | | |
| 3. Expand | Conduct a literature review of articles published in the last three years to update literature in Chapter 2 of the dissertation. Expand literature review to include studies from other perspectives, including the global south. | Use as the basis for journal article. Writing in a global context will help me to compete for academic positions in other countries. | Look for a journal with a global audience. Also, look for call for chapters in books from top-tier publishers that might have broader distribution. |
| 4. Adapt | | | |
| 5. Apply Combination of options | | | |

analysing, and incorporating new literature or collecting new data in a follow-up study. Begin developing a more detailed set of goals and outlining the steps you will take to accomplish them.

## *Exercise 3: start outlining your publication strategy*

Building on the Chapters 1 and 2 Part 1 exercises, create a preliminary outline for your publication strategy. You will continue to refine this outline, and add new thoughts and details, as you work through the exercises suggested in this book. You will also want to set up a file (paper, electronic, or both) where you can collect important resources that support your publication strategy.

Here is a suggested outline that you can amend to suit your needs. Add or delete sections to personalise your publication strategy. (Go to the Routledge book website or www.path2publishing. com to download an outline you can fill in.)

  i **Goals: What goals are important to you?**
   a  Career goals
   b  Impact, social change, social justice goals
   c  Personal/life goals
   d  Other goals
  ii **Content: Given the assets you identified in Chapter 1, what content do you have**

**and what new content will need to be developed?**

a  Existing research
b  New research
c  Literature
d  Theoretical propositions
e  Findings or results
 f  Recommendations or implications
g  Experience
h  Observations
 i  Other

iii  **Publication Types: Given what you know at this point, what types of publications are you considering?** As you move through this book, you will explore a variety of conventional and emerging ways to publish your work, so you can continue to add the ones that make sense for you.

iv  **Strategy for Each Type of Publication: Begin making notes about your strategies and add as you continue to work through the coming chapters.** Add as many types as you want to work through in order to determine your priorities.

a  Type 1
   i  Content
      1  Adapt existing
         a  Steps
         b  Timeline

       2 Create new
        a Steps
        b Timeline
    ii Sole author or co-author?
       1 Agreement(s)    established    with co-author(s)
   iii Respond to call for papers or submit proposal directly to the editor
   iv Submission steps and protocols
    v Writing
   vi Revising and finalising the manuscript
  vii Submitting the manuscript
 viii Addressing review comments
   ix Promoting and/or presenting the work
    x Timeline

## *Exercise 4: consider relevant ethical issues*

- Locate and familiarise yourself with at least one code of ethics relevant to your field. Also, check out the website of the Committee on Publication Ethics at https://publicationethics.org/.
- Identify the ethical issues you will need to consider when publishing in your field. You might want to discuss with other writers and/ or record your thoughts in your writer's journal.
  a Ethical issues in respect of other scholars
  b Ethical issues in respect of participants

   c  Ethical issues in respect of my readers
   d  Ethical issues in respect of where I publish
   e  Ethical issues related to self-plagiarism
   f  Any other ethical issues specific to my field

## Good practice points

- At the initial stage, keep decisions open about how each asset might be used or how it might fit with a particular type of publication.
- Keep in mind that the same asset might be used for different purposes, depending on the types of publication.
- Plan to format and organise your publication strategy in a way that allows for it to be a living document with flexibility to change as you progress through your plans.
- Choose the publication strategy format that works best for you: paper or electronic.

## References

Kara, H. (2015). *Creative research methods in the social sciences: a practical guide*. Bristol, England: Policy Press.

Kara, H. (2017). *Research and evaluation for busy students and practitioners: A time-saving guide* (2nd ed.). Bristol, England: Policy Press.

Luey, B. (2010). *Handbook for academic authors* (5th ed.). Cambridge, MA: Cambridge University Press.

Rallis, S. F., & Rossman, G. B. (2012). *The research journey: Introduction to inquiry*. New York, NY: Guilford Press.

Roberts, C., & Hyatt, L. (2018). *The dissertation journey: A practical and comprehensive guide to planning, writing, and defending your dissertation* (3rd ed.). Thousand Oaks, California: Corwin Press.

Salmons, J. (2007). Expect originality! Using taxonomies to structure assignments that support original work. In T. Roberts (Ed.), *Student plagiarism in an online world: Problems and solutions*. Hershey, NY: IGI Reference.

Salmons, J. (2010). *Online interviews in real time*. Thousand Oaks, CA: Sage Publications, Inc.

# 3   Why, when, and how should I publish journal articles?

**After studying this chapter, you will be able to:**

- Explain the publication process for scholarly journal articles.
- Recognise legitimate or predatory journals.
- Analyse ways to use your thesis, dissertation, or academic writings as the basis for journal articles.
- Evaluate implications for your publication strategy.

## Overview

No doubt you have read many journal articles during your postgraduate studies, but have you ever considered writing one? There are tens of thousands of academic journals in the world, on a vast range of subjects, and in many languages. Scholars share articles that develop propositions, theories, and discoveries made in the past, or reject them. If you're going to write a journal article, you will be taking part in this ongoing international conversation. Keeping the suggestions from this chapter in mind can help you to situate your work appropriately among the other contributors to the scholarly conversations going on in your field.

While major publishers have divisions dedicated to journals, professional associations and industry groups also distribute journals. Some are purchased by subscription and some are available to anyone online. You have probably accessed subscription-based journals through your college or university library. As an alumnus or alumna, you may have restricted access, or no access, to your library's database of journals.

Articles published in a *scholarly* journal are reviewed by an editor and experts in the field through a multistage process known as peer review. Articles published in a *professional* journal are typically reviewed by one or more editors. Ideally, articles in academic journals form the basis of a worldwide conversation between academics that develops over time (Thomson & Kamler, 2013, pp. 56–59), and articles in professional journals form the basis for conversation about improvements in policy or practice. This is inherently challenging, partly because there are so many journals in so many languages. On the other hand, sometimes special issues of journals are dedicated to specific topics, perhaps following on from discussions at a conference or other event, or defined by an editorial board. These can be very useful in collecting and developing expert thought.

Even so, the majority of journal articles are honest, credible contributions to knowledge. Getting your research published in a scholarly journal can establish you as someone who has successfully carried out research, or has insights into problems, research methods, or theories. Yet writing journal articles and getting them published can seem like one of the dark arts of academia to the uninitiated. However, like so many things, it's not as difficult once you know how.

This chapter will take you through the process of developing a journal article from your thesis or dissertation. We will look at how to find a suitable journal, how to avoid desperate and predatory journals, how to plan and write your article, and how to deal with reviewers' comments.

## How to use your identified assets in journal articles

Use your identified assets to create academic journal articles (Figure 3.1). Do you need to use one or more of these approaches?

- To extract, take out a section of your thesis, dissertation, or other scholarly writings and use it as the basis for a journal article.

# How can I use assets when writing journal articles?

**Extract**
- Extract sections and use as the basis of 1 or more articles.

**Condense**
- Distill lengthy writing to fit requirements for articles.

**Expand**
- Clarify points, add examples, update literature, or build on the findings and analysis to create engaging article(s).

**Adapt**
- Focus on aspects relevant to new audience, explain terms or concepts.

**Apply**
- Offer recommendations for using findings in scholarship or practice.

*Figure 3.1* How can I use assets when writing journal articles?

- To condense, distil key elements of lengthy doctoral writings into a journal article. Given the difference in length, what appears as one or more pages in the thesis or dissertation might need to be presented as one or two paragraphs.
- To expand on your assets, add elements that will strengthen the original writing. Depending on how long it has been since you completed your doctoral studies, you might need to include new literature or current descriptions of research problems to bring it up-to-date.
- To adapt, use your existing writings for a new purpose or audience. If you want to publish in a journal outside of your original discipline, or write for a global audience, you might need to adapt your descriptions or define terms.
- Some journals want articles that readers can use. If that is the case, you might want to explain how to apply ideas in practice, or in future research.

## How to find suitable journals

You will be familiar with major journals in your discipline or field of study through assigned readings in your coursework or recommendations from your doctoral supervisor. Respected journals are widely cited and attract articles from prominent senior scholars. They can also be extremely competitive for new contributors, so you might want to look more widely for opportunities that will allow you to get started.

If you want to find additional journals that might welcome your article, look strategically for those that focus on your topic or your methods. Scan the reference lists of articles written by leaders in your field, and look up any journals that are new to you. Look for specialised journals published by professional societies or associations in your field, discipline, or research interest. Make sure any journal you submit to has a suitable focus and readership for your work (Schimel, 2012, p. 192).

One way to broaden the potential journals is to look beyond your discipline. Your topic can lead you to journals in related fields. For example, our PIP Nathan will be interested in writing for journals in psychiatry. However, he should also consider writing for journals in sociology, because of his focus on families, and in military studies, because that is another aspect of his research.

You might want to write an article focusing on the research approach used in your study. So, if you have developed a new method for conducting interviews, look for journals specialising in research methods. Equally, if you want to write an article on an unusual ethical dilemma, you may want to find a journal specialising in research ethics.

You will need to consider whether to write for a subscription-based or an open-access journal. The decision is likely to depend on the interplay between cost and budget, the importance for you of a journal's impact factor, and the prestige of a given journal in your field. You will also want to consider the target audience for your article(s). Are you aiming to reach academics and students who can read journals in their college or university library, or are you hoping to reach people who are not associated with educational institutions?

## When a journal finds you

Sometimes an academic journal will come to you. There are four main types of people who may ask you to write a journal article:

1 A legitimate journal editor
2 An editor of a special issue of a legitimate journal
3 A desperate journal representative
4 A predatory journal representative

The editor of a legitimate journal may hear you speak at a conference or hear about your work from a colleague. They will approach you in person or by email to ask if you're interested in writing an article for their journal. You will be able to discuss the article with them and come up with an angle that you can write and they want to publish. Publication is not guaranteed – you still have to write a good article – but it is much more likely than when you approach a journal yourself.

The editor of a special issue will also know of your work and will approach you to ask if you want to contribute. These themed issues and the articles they contain tend to be more widely downloaded and read than the more general issues of journals. This means that publishing in special issues can be a useful way to increase your visibility. Again, publication is likely as long as you do a good job with the article.

Desperate journals are a different matter. They will approach you by email, which may, at first sight, seem very flattering. The email will praise your previous work, or your expertise, or both, then invite you to write an article for their journal, or even to edit a special issue. But when you look more closely, these emails usually seem a bit odd. They may ask you to write in areas where you have little knowledge and no expertise, or to take on onerous editing responsibilities for little or no reward. The language can be surreal and the timescales are usually unrealistic, such as three weeks to produce a journal article.

Emails from predatory journals often look very similar to those from desperate journals.

The difference is they will charge you a fee for publishing in their journal, which may not even exist. Some legitimate open-access journals apply charges for publishing as a way of financing their business model, so it can be difficult to tell when a journal is predatory. A couple of good ways to check include looking up the journal editor online; they should have significant relevant academic expertise. Also, the editorial board should be listed, and should consist of people who have similar expertise to the editor – and are both real and still alive. The journal's website should look professional, with links that work appropriately, and should have an email address linked with an academic institution or the journal itself. A legitimate journal will never ask for any payment before an article is accepted for publication. If you're still unsure, check with an academic librarian, or the website https://thinkchecksubmit.org. Remember, if it looks too good to be true, it probably is.

### How to research a journal's submission protocols

Once you have identified one or more suitable journals, and made sure they are not desperate or predatory, it is time to research them in more detail to determine which one(s) are relevant to your work. First, find their web page and download their guidelines for authors. These should include useful information like the journal's word count and its submission method. Second, if you're not already very familiar with the journal, search it and download some articles. Look for commonalities. For example, do they all have a title with two sections separated by a colon? Do they all have a two-paragraph conclusion? If you spot any such patterns, make a note and ensure your own article follows the same format. Also, you should cite at least two articles from the journal in your own article. These steps will increase your article's likelihood of acceptance.

It can be useful to prepare a table comparing the different journals. Here is an example of a table (Table 3.1) comparing some prominent journals in the field of qualitative research methods.

This exercise illustrates the point made above that you can't always find relevant journals by using intuitive keywords. For example, the *International Journal of Social Research Methodology* doesn't include the word 'qualitative' in its title because it publishes both qualitative and quantitative research.

We suggest that you date your table because journals change over time. Therefore, if you want to publish another article in a similar journal at a later date, you will need to update your table.

**Table 3.1** Journal review example

| Journal | Impact Factor (2018) | Cost to Publish | Word Count | Abstract | Keywords |
|---|---|---|---|---|---|
| **International Journal of Qualitative Methods** | 1.387 | CAN$1000 (CAN$375 for students) | 3,500–7,500 excluding references and abstract | Not stated | Not stated |
| **Forum: Qualitative Social Research** | Not stated | £0 | Not stated | 100–200 words | 5–10 |
| **International Journal of Social Research Methodology** | 1.394 | n/a | 8,000 maximum including everything | 150 words | 4–5 |
| **Qualitative Research** | 2.951 | n/a | 8,000 maximum including everything | 100–150 words | Up to 10 |

It will be interesting to see the changes since your initial investigation. Also, it is interesting to see the variations between journals. In the table above, the first two are open access, the last two are subscription-based. With subscription-based journals, if you have a budget, you can usually pay for your article to be published as open access. However, it is optional, unlike the costs for open-access journals such as the *International Journal of Qualitative Methods.* Having said that, some open-access journals will waive charges in certain circumstances, such as for independent, unemployed, or retired authors.

## How to plan a journal article

When you have decided on a journal to write for in the first instance, and – crucially – you know the maximum word count (and sometimes the minimum), you can start to plan. If the journal you have chosen for your first submission doesn't give a word count, don't take that as licence to write 15,000 words. Check the word counts of articles within the journal or related journals and use them as a guideline.

Download and read the author guidelines for your intended journal. Make a note of any requirements that may be relevant for you. Look at specific dates, as well as the timing and expectations at each stage of the submission process. You will notice that some journal editors ask for the completed article to be submitted, and others ask for an abstract and description. In the first case, you will receive feedback on the entire article from the editor and peer reviewers. In the second case, you will receive initial feedback from the editor.

Create a working title: This is a title for now, not necessarily the final title. A working title is useful for reference to help you focus on the premise of the article. Then list the key points you want to make and link them with the literature you want to cite. Are there any gaps in your reading? If nothing else, there is likely to be a time gap since you wrote your thesis or dissertation. Make a brief note of your gaps and think about how you will fill them.

Your next job is to create a frame for your article. Include the headings you are likely to use, the approximate word count for each section, and a brief description of each section's intended content. This is not to tie you down but to provide a structure to support your work (Murray, 2009, p. 122). Any or all of it can be changed as you go. Yet it is worth creating, because this structure, together with your working title and notes, will enable you to write your article more easily and efficiently than you could do otherwise.

Once you have your working title, brief notes, and draft structure, you are ready to start work.

## How to write a journal article

One of the great advantages of good planning, as described in the previous section, is that you can use small amounts of time to move forward with your article (Murray, 2009, p. 71). In just five minutes you can do a quick literature search or write a couple of sentences. In 15 minutes, you can read part of an article or write a paragraph, maybe even two. In half an hour … you get the idea.

Don't aim for perfection in your first draft, just get the words down. Remember the first draft of any written work is to tell yourself the story, the second is to tell the story to other people. Editing can wait until you have all the basics down and all the citations you think you need. Then go through the article again to make sure that it contains everything you want to say, the narrative flows, and the text fits within the maximum word allowance. Check for any repetition or waffle, and make sure every word in your article counts.

Because journals are published electronically, they can include embedded media, audio, or links to original datasets. If you want to include any non-prose items, such as figures, tables, diagrams or illustrations, or links to media, be sure to follow the author guidelines for your intended journal.

When you have made your draft article as good as you can, leave it alone for at least a week so you can come back to it with fresh eyes. Use this time to solicit feedback from one or two trusted colleagues.

### Seeking and working with pre-submission feedback

It is really important to get, and use, some feedback on your article before you submit it to a journal (Nygaard, 2015, p. 143). Writing is difficult and none of us can do it as well alone as we can with input from others.

Having said that, feedback can also be hard to take (Nygaard, 2015, p. 145), even if it is entirely constructive. Remember that the person giving the feedback is trying to help you. If acceptance doesn't come naturally, you will need to develop skills in accepting feedback graciously. After all, when your work is published, it will be available for everyone to critique. Learning from research is an iterative process so your conclusions are likely to receive criticism sooner or later.

For now, though, it is worth thinking carefully about who to ask for feedback: someone who knows a lot about your work or someone who knows a little? (Nygaard, 2015, p. 144). Someone who knows your work well may be able to give more detailed feedback, but on the other hand they might want to influence your approach more than would be ideal. Someone who doesn't know your work so well may bring a fresher and more balanced perspective, but their feedback could be less nuanced. Either way, you are asking a favour, so bear in mind the person's workload and other commitments, and perhaps what you can do for them in return.

Try to get feedback in writing, and give yourself time to consider and absorb it before you respond (Kara, 2015, p. 124). When you're ready to work with your feedback it can be useful to take an analytical approach. Use a bullet pointed list, in your own words, to keep track of requested revisions.

There is more information relevant to dealing with feedback in the section on dealing with reviewers' comments.

## How to submit a journal article

Check through the author guidelines for your desired journal one more time and make absolutely sure your article complies with the journal's requirements. When you've done a lot of work on a journal article, it feels incredibly depressing to have it rejected because, for example, you've got the references in the wrong format.

Some journals ask you to email your submission directly to the editor, others use online submission systems such as ScholarOne or Manuscript Central. These are exacting and fiddly to use. Set aside a couple of hours for submission, particularly if you haven't used an online submission system before. The systems are not actually difficult to use, just tedious and time-consuming. You will have to work through several webpages requiring you to provide a range of information, from your own biographical details and those of any co-authors, to the names and contact information of potential reviewers, before you can upload your draft article. Some journals want figures and diagrams to be uploaded separately, and this will involve extra work if you have included them in the body of your text.

You must make sure your draft article is anonymous for submission, in order to prepare for the blind review. This means removing your name and the names of any co-authors, and also removing any other parts of the article that could identify one or more of the authors; you can mark these as 'redacted' if necessary. Then you need to digitally anonymise the article. In Word (at the time of writing) you do this with the document open by clicking File, then Check for Issues, then Inspect

Document. In the dialog box, select Document Properties and Personal Information. (Make sure nothing else is selected or you could end up deleting any tracked changes or comments.) Then click Inspect. If anything is found, click Remove All. You can then Reinspect if you want to make sure, or simply close the dialog box.

## Dealing with reviewers' comments

As when dealing with pre-submission feedback (see above) first you need to give yourself time to react. Whether it's the exultation of an acceptance, the misery of a rejection, or the mixed feelings that come with requests for revisions, you need time to process your emotional response before you do anything else.

Because of negativity bias, negative comments – even when constructively phrased – have more impact on most people than positive comments. We need to work to counteract this bias. So, unless you've received very favourable comments and you want to revel in their glory, we recommend waiting at least 24 hours before you read the comments again. This can help you to take a more balanced view, which is useful because if it's a rejection or revisions, you'll need to see how your work can benefit from the reviewers' input before you send it off again (Murray, 2009, p. 188). Murray (2009, p. 196) suggests this can be quite a challenge, especially if the reviewers have different views of your work and how it can be improved. Your journal or special edition editor may offer some guidance and if so, you should take that into account. But sometimes they leave it all to you.

If you don't understand (or aren't sure you understand) a reviewer's comments, seek a second opinion from someone more experienced (Thomson & Kamler, 2013, p. 139). Once you are clear about what they mean, we suggest you treat the reviewers' comments as data for analysis. Create a table with one column for the comments and another column for each reviewer. Then enter each substantive comment into the first column and put a mark in the other columns for each reviewer who has made a similar point. This helps you to pick up the instances where reviewers are effectively saying the same thing though perhaps in very different ways. It also shows you which comments have been made by all or some reviewers, and which only by one of the reviewers.

Reviewers' comments come in three categories:

1  The no-brainer (act on this)
2  The no-thanks (don't act on this)
3  The oh-wait (probably act on this, though not necessarily in the way the reviewer suggests)

Your next job is to sort comments into these categories. If a comment has been made by more than one reviewer you should take it more seriously. That doesn't mean you have to implement it, but if a comment has been made by all reviewers you will need an exceptionally good reason not to implement it. If a comment has been made by only one reviewer, that in itself might be a reason to decide not to implement it, though you should also have at least one other reason.

Once you have sorted the comments into their categories, you can list them by category in the first column of a new table with three further columns: one for the no-brainers, one for the oh-waits, and one to help you formulate the point-by-point response to reviewers' comments that most journals require. Then you can note down what you plan to do in response to each comment and what you plan to say to reviewers. This is useful because you can dip into it when you have a spare half hour or so, and find a job or two to do to bring you closer to the finish line. As you revise the text of your article, use tracked changes or coloured highlighting so an editor can easily see where you've made amendments.

It is important to be polite in your response to reviewers' comments, including the no-thanks, even if you think they're utterly ridiculous. Some reviewers' suggestions seem to be based more on what they would have written than on what you actually have written and this can be quite annoying at times. When you come across a suggestion you really don't want to implement, there are some tactful ways to say so, such as:

- 'This is an excellent suggestion though unfortunately beyond the scope of this particular project'.
- 'I can see how this suggestion would improve my work but sadly I am unable to incorporate it within the allocated word count'.
- 'This is a really interesting idea. I have considered it carefully and concluded that it doesn't quite fit with the thrust of my current article, but it will influence my thinking for future projects'.

Remember you are the author and, as such, you have authority. While authors do need constructive input from reviewers, and your work should benefit from intelligent use of their feedback, you don't have to do everything a reviewer says (Nygaard, 2015, p. 148).

## Managing rejection

Receiving a rejection of your work can be upsetting. Practice self-care and give yourself time to recover. Not too much time, though, because you have more work to do. Remember, a rejection is only a rejection from *this* journal. It doesn't mean your work is worthless; sometimes it's only because they already have plans to publish something similar.

A desk rejection (from the editor) may not come with any feedback. However, most rejections have at least some feedback attached. *It is crucially important to use this feedback*. Feedback from editors and/or reviewers is the valuable information that can help you to get your work published. Consider it just as you would if you'd received a 'revise and resubmit' response, using the methods set out above. Also, identify another journal that might be suitable for your work. Research its submission protocols, as outlined earlier in this chapter. Then figure out what (if anything) you need to change in your manuscript, based on the new submission protocols and the feedback you have received. Make all the amendments that seem sensible to you, check carefully for any errors, and get your article back out into the world.

## PIPs

As **Kris** wants to be an academic, journal articles are essential to being hired for a faculty position. Journal articles are particularly important in some disciplines or in educational settings where there is an audit culture to assess whether faculty members have met set benchmarks or criteria. To merit promotion or tenure, Kris not only needs to publish journal articles, she also needs to publish in journals with a high impact factor, and to keep publishing several journal articles each year. Kris really struggled with her first journal article. There seemed to be so much to think about! But when she had submitted one, the prospect of doing another seemed easier.

**Ella** is in a different position. Publishing some articles in reputable journals is likely to mean that higher education institutions take her more seriously. She doesn't need to publish in scholarly journals as much or as often as Kris, so can also choose other types of publications to further her career. She is delighted to have persuaded her doctoral supervisor to co-write a journal article based on her research findings. She plans to extract the description of the research problem, condense her literature review, and adapt the findings described in her thesis to meet the journal's focus. Her supervisor will guide her through the process, and Ella thinks publication will be more likely as a result of their collaboration.

**Nathan** does not need to publish scholarly journal articles at all because he is aiming to develop his career as a practice-based researcher. He chose to translate his research findings into a practical how-to article for a professional journal that is well respected in his field. Nathan investigated ways to pitch ideas to this journal and was pleased to discover that the process was quite straightforward. The journal editor was interested in his work and happy to publish his article.

## Chapter summary

Scholarly journal articles are probably one of the most difficult types of publications due to the rigorous peer review process. The more prestigious the journal, the more competitive it is for submissions by scholars across the world. This process can seem intimidating because no one likes to be the object of critical review by strangers. By doing due diligence and selecting appropriate journals, you are taking important first steps. Carefully analysing your identified assets will help you to feel more confident about the quality of the article you submit.

## Exercises and questions for reflection: planning and writing a journal article

Find templates and worksheets for this chapter's activities on the Routledge book website or www.path2publishing.com.

**Reflection**

When you think about the journal articles you want to write, what purpose do you hope they will achieve?

**Exercises**

*Exercise 1: research options and make plans*

1  Find three or four academic journals relevant to your topic, findings, methods, and/or ethical approaches. Make a table with brief notes of their key requirements.
2  Make initial plans for a journal article for one of these journals, using the following headings:
    a  Journal name
    b  Word count
    c  Any other relevant requirements
    d  Working title for article
    e  Key points to make
    f  Relevant literature (already read)
    g  Relevant literature (need to read)
    h  Search terms to use to find literature
    i  Timescale/deadlines

## Exercise 2: add journals to your publication strategy

Update your publication strategy you began to outline in Chapter 2 to include goals and key steps associated with publishing one or more articles.

## Exercise 3: consider the value of journals in your field or discipline

The readership and use of journals may influence what articles you choose to write. Look at the most-read and most-cited articles in leading journals in your field. Are they intended for use as assigned readings in academic courses, or for ongoing edification of scholars? Do they serve as discussion springboards at conferences, as the basis for policy changes, or do they serve other purposes?

## Good practice points

- Research your desired journal thoroughly
- Write the first draft, edit the next
- Stay within the maximum word count
- Follow author guidelines
- Analyse and organise review comments
- Re-submit

## References

Kara, H. (2015). *Creative research methods in the social sciences: A practical guide*. Bristol, England: Policy Press.

Murray, R. (2009). *Writing for academic journals*. Maidenhead, England: Open University Press.

Nygaard, L. P. (2015). *Writing for scholars: A practical guide to making sense and being heard* (2nd ed.). London, England: SAGE Publications.

Schimel, J. (2012). *How to write papers that get cited and proposals that get funded*. Oxford, England: Oxford University Press.

Thomson, P., & Kamler, B. (2013). *Writing for peer reviewed journals: Strategies for getting published*. Abingdon, England: Routledge.

# 4 Why, when, and how should I publish books?

**After studying this chapter, you will be able to:**

- Identify motivations and options for publishing books.
- Define types of book formats and steps for publishing each one.
- Understand options for using your identified assets as the basis for a book or books.
- Evaluate implications for your publication strategy.

## Overview

A generation ago, academic and professional books had a fairly standard format: Mostly they would be printed paper books containing 75,000–80,000 words. Each book would contain a table of contents, a bibliography, an index, and perhaps a few appendices. Of course, the format and length varied slightly at times, but generally people knew what to expect. Now, books can be delivered in print, electronic, and/or audio versions, at a wide range of lengths. There are many options, so it is important to consider which will further advance you toward your career goals.

Some people are still harbouring the misconception that publishing a book will make you rich. This does happen, but only to a very few people, mostly novelists or celebrities. Writing related to academia is highly unlikely to make you rich. Here's an example. Helen Kara published her first book in 2012 and her second in 2015. They both sell quite well. Her 2017–2018 royalties for those books were £947. That's a nice sum of money to receive, but it's not a 'getting rich' sum of money. Janet Salmons isn't getting rich either but receives a bit more in royalties from six books in print at the time of writing. She has increased royalties by allowing two books to be distributed in an electronic database. If we thought in terms of an hourly rate for the time we spend writing … well, it would make us cry.

What books *are* good for, though, is boosting your career. As a result of Helen's second book, she has had work overseas and now regularly teaches in universities and gives keynote speeches at conferences. As a result of Janet's collection of books, she is invited to give presentations, workshops, and seminars, and to write for various publications.

Books opened the doors for other assignments that do pay well and offer credibility as experts in our respective fields. But writing a book is a difficult, time-consuming, long-term project. To write a book, you need to love writing, be self-disciplined, and be able to sustain a project over several months, maybe even years. If you have completed a dissertation or thesis, you have a pretty good idea of the kind of commitment needed to complete a significant piece of writing. Book publishing, too, is very much about taking the long view. This chapter will lead you through the steps of the process.

## How to use your identified assets in books

Use your identified assets to create books (Figure 4.1). See Chapter 5 for more suggestions to use in chapters for the book. Consider using one or more of these approaches:

- When you look at your assets, are there important concepts, definitions, or other sections to *extract* and build up?
- Are there sections that you can *condense* to highlight relevant points?
- Where will you need to *expand* on your assets? This is a particular consideration if you have

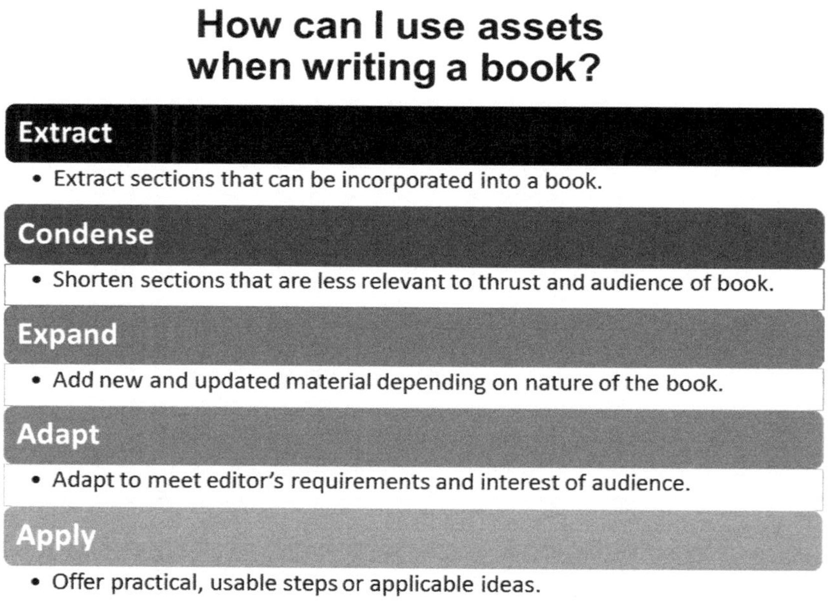

# How can I use assets
# when writing a book?

**Extract**
- Extract sections that can be incorporated into a book.

**Condense**
- Shorten sections that are less relevant to thrust and audience of book.

**Expand**
- Add new and updated material depending on nature of the book.

**Adapt**
- Adapt to meet editor's requirements and interest of audience.

**Apply**
- Offer practical, usable steps or applicable ideas.

*Figure 4.1* How can I use assets when writing books?

been out of your doctoral programme for a while and need to add current references or research.
- What *adaptation* will help reach the intended readership for the book?
- What recommendations can you offer to readers who want to *apply* findings and insights?

## Book proposals

If you want to write or edit a book, the first thing you have to write is a proposal for that book. With non-fiction books, including academic books, you don't need to write the whole book before approaching publishers. Most publishing contracts are awarded on the basis of a proposal alone. However, if you have no track record as a professional author or you are writing a particularly unusual book, the publisher and reviewers may also want to see a draft chapter.

The format for a book proposal varies between publishers, but there are some elements that are common to all proposals. Any publisher will want you to conduct a competitor analysis to demonstrate what else has been published in the field and show that the book you are proposing fills a gap. They will want to be sure that they are publishing an original book, not the one that is too similar to a book already in existence.

A publisher will also want a brief description of the book's concept and a detailed chapter-by-chapter explanation of the proposed content. They will want your view of the target audience or audiences, and suggestions of people who might review your proposal and (if applicable) draft chapter. Many publishers also want to see whether you have a network, an online presence, or a reputation that will help to bring attention to the book.

Some publishers will prefer that you make an initial contact with the commissioning or acquisitions editor before you send in the book proposal. This is worth doing even if the publisher doesn't encourage it, to find out whether they already have a similar book in the pipeline. If they do, they are very unlikely to want to publish yours, and discovering that up front can save you spending days on a pointless proposal.

Information about publishers' submission processes and associated requirements will be posted on their websites. The processes can vary somewhat, but again there are some commonalities. When the commissioning or acquisitions editor is interested in your proposal, it usually goes through several more steps before formal acceptance, such as some kind of review and an internal acquisitions meeting.

For an academic book, the proposal is likely to go through a peer-review process. For textbooks, publishers may ask faculty members who teach in the appropriate field to look at the proposal in light of their own instructional needs. You may be asked to revise your proposal on the basis of feedback from these people.

Once your book has made it through all of these hoops, you will be issued a publishing contract. See Chapter 7 for more information about what to do at this stage.

## Doing research for a book

Whatever kind of book you write, you most likely need to expand on the research conducted for your doctoral programme. How thorough book research should be varies between authors and books, and between publishers and editors. On the practical side, you or a collaborator will need to research the market and existing books to put together a competitor analysis as required for book proposals. On a more substantive side, you might need to update the literature or data, expand on background research to strengthen your case for the book, or adapt work conducted in a specific setting to reflect a larger context.

Depending on the nature of the book, the field, and the time that has elapsed since your doctoral studies, book research can be informal or formal. If more than two years have passed since you collected the data, you should consider whether changes that have occurred during that time mean you need to collect and analyse new data. A series of brief or in-depth interviews, or an online survey, could expand on your doctoral research and add colour and veracity. Depending on the nature and sensitivity of the topics and the organisational or institutional setting, a simple consent agreement might suffice or a full ethics review from an academic institution or agency could be required.

In other cases, informal desk research such as finding current facts and figures to bolster your argument, reading current literature, or observing public situations could be adequate. This kind of research doesn't need to go through ethical approval: It simply needs to be thorough enough to satisfy your editor and readers.

## Sole-authored books

You are responsible for all of the content in a sole-authored book. The sole-authored book is the largest, and most substantial publication you are ever likely to produce. Your dissertation or thesis was a significant piece of writing, equivalent to a book in length and complexity. However, very few people are able to publish their dissertation or thesis as it stands, or with only minor tweaks. Think about how your findings might apply in the wider world. You will most likely need to remove or rewrite some of the more doctoral-specific elements, such as your literature review and methodology chapters, and expand on your findings and conclusions. You may find useful materials to help you from your other identified assets. In doing this, you are adapting your work to an audience beyond your own institution and field.

Being clear about your target audience or audiences is essential, not only for marketing purposes when the book is done but also for helping you write it effectively. Think about the needs of your audience(s). What kind of language would suit them best? How can you help them navigate through your text? Would it be helpful to include pedagogical features? (See below for more on these.) Are there visual elements you could usefully include such as photographs, diagrams, maps, or charts? Does your book need a glossary?

When your plans are clear and you are ready to start writing, set yourself a weekly goal for the first draft. Helen Kara finds that 2,500 words per week works well for her. That's 500 words per day Monday–Friday (and 500 words is approximately one page of typed A4 in Arial 12-point). Though it must be said that sometimes, in practice, it's 2,500 words on a Sunday afternoon.

Janet Salmons sets goals in relation to completion of specific stages of writing projects, regardless of word count. Different approaches work better for different people; another option is allocated time slots. However you find you work best, the highest priority is to keep the words coming regularly. A book is so long that if you leave it all until the last minute – or even the last month – you will miss your contractual deadline.

## Co-authored books

If you are reluctant to take full responsibility for a book, consider writing with co-authors. Co-authors can bring additional insights to the book and perhaps balance and complement the work you have to offer. Meetings and conferences of professional societies or associations can be good places to find potential co-authors. Look for others who share your interests, and other researchers who have studied similar questions from different angles.

A more experienced co-author may be able to help you navigate the opaque world of publishing. If you want to find a co-author, look for someone who is willing to negotiate a fair agreement about who does what. This is not only in terms of the contribution of content to the book but also in terms of the other logistical and administrative aspects of the work such as contract management, deadline management, marketing, and so on.

The keys to successful co-authoring are clear and regular communication, achievable deadlines, and good time management. You also need to be able to give and receive feedback in a professional manner. Everyone has different working styles. It's well worth discussing your respective working styles with any potential co-author, to make sure you can work well together before you start on a project. You can learn more about collaborative writing from another book in this series, *Collaborate to Succeed in Higher Education and Beyond: A Practical Guide for Doctoral Students and Early Career Researchers* (Lemon & Salmons, 2020).

## Edited collections

If you would like to publish your work in a book without producing a full-length manuscript, you could consider editing a collection. This option would allow you to publish one or more of your own chapters together with the work of other researchers in your field.

An edited collection is a book-length compilation of chapters by different authors (Salmons, 2017a). An edited book collection is typically organised around a central theme or focus, and as editor, you determine the theme and content. While some editors anthologise previously published materials, others commission new, original chapters. Davis and Blossey (2011) observe that edited collections are particularly useful in emerging fields of study by pointing to new research ideas, experimental designs, or analysis. Nederman (2005) suggested that at its best, the multi-authored, edited volume can promote fruitful exchange of ideas in a way that would not occur in research monographs or journal articles. When you contemplate your potential edited book, it is important to consider how you will synthesise ideas into a coherent whole and seed new thinking in your field.

Given these responsibilities, the skills needed to edit a book are different from those needed to write a book. As a sole author, you are responsible for the vision, plan, and for writing all the book's content. For an edited book, you are responsible for the vision, plan, and management of others' content (Salmons, 2017a). Editorial responsibilities include selecting a publisher, developing a proposal, submitting it, and negotiating the contract (see Chapter 5 for more on this). The editor is also responsible for developing inclusion criteria for the book, crafting a call for chapters, distributing the call to potential contributors, and/or inviting other writers to contribute. Editors also spend a lot of their time chasing contributors and reviewers.

Depending on the publisher's requirements, the editor may also need to assemble an editorial board for the book. Typically, the editorial board commits to some responsibilities for the review process. Another option is to ask chapter authors to review other draft chapters intended for the book.

If this option appeals to you, you may want to invite a colleague to serve as a co-editor and contributor. This would allow you to share the considerable work involved in putting together an edited collection. Having two of you on the job is also helpful if you need to deliver bad news to a potential contributor. For example, saying, 'our view is that your chapter, as it stands, doesn't fit the brief we gave you' is a lot easier than saying 'I think your chapter doesn't fit' because the contributor is less likely to take it personally.

There is a lot of work in an edited collection, but this is unlikely to be more work than would be involved in writing or co-writing a whole book. Also, edited collections can help you to build your network of like-minded people as you work with your contributors.

## Short books

Many publishers are now releasing short books of various kinds. These may be as short as 5,000 words or as long as 50,000 – the same length as some theses or dissertations. Very short books are often distributed as e-books alone, but otherwise they may be distributed as print books or e-books or both. Short textbooks may be interactive, with exercises devised for students to write their answers into the book.

Short books can often be produced more quickly than a full-length book or a journal article. This may be an attractive option if you have material that extends beyond the length of a single chapter or timely research you would like to disseminate quickly, and/or you would like your material to be available as a standalone book.

A short book may allow you to select and extract elements of your doctoral work, edit or format as needed, and get the work published. Otherwise, the only respect in which short books are easier to write is that there are fewer words for you to produce. You will still need to create a full proposal, think about your audience(s) and their needs, and write well.

Academic short-form book proposals are likely to be subject to peer review, and short textbook proposals may be sent to faculty members for feedback. As with full-length books, you can write short books alone or with one or more co-authors. The book you are reading now is a co-authored short book.

Academic books may contain pedagogical features. 'Pedagogical' means 'relating to education', and pedagogical features are aspects of a book designed to help readers learn. They can be used to explain difficult concepts or procedures, or to offer the opportunity for the student to practice skills in realistic exercises (Salmons, 2017b). The PIPs, exercises, and good practice points in this book are examples of pedagogical features. Other types of materials can include:

- Case studies
- Vignettes
- Discussion questions
- Further reading suggestions
- Chapter summaries
- Checklists
- Assignment suggestions
- Sample syllabi
- Slide decks

Pedagogical features can be made available on a book-related website maintained by the publisher and/or the author. While using the publisher's site means purchasers will readily see the materials, creating your own site has advantages as well:

> When we accompany the book with our own website or blog we can make fresh content available to instructors who have adopted a book. When we give presentations, or try out new applications of the ideas from the book, we can make them available for others to use. As authors we can interact directly with instructors, hear about their dilemmas – and develop materials to help. (Salmons, 2017b)

Well-thought-through pedagogical features add value, particularly to textbooks, guides, or manuals. They provide instructors, workshop facilitators, training professionals, and students with the tools they need to adapt and apply research findings for their own purposes.

## Book completion after submission

Commissioning or acquisitions editors are part of the editorial department that oversees the initial stages of a book's creation. The editorial department is responsible for commissioning or acquiring books, negotiating contract terms, and arranging for peer reviews, cover design, and an index. Control over the final book also varies from one publisher to another. Publishers with the least control focus on simply printing and distributing the book, while publishers with high levels of control might recommend changes to writing style, content, sequence of chapters, font, placement of figures, and/or cover design (Thomas & Brubaker, 2008).

Other departments involved with getting the book produced are:

- Production: responsible for turning the final manuscript into an actual book, including page layout and formatting e-books
- Marketing: responsible for promoting and publicising books
- Sales: responsible for liaising with booksellers and event organisers and getting boxes of books to wherever they need to be

In some publishing firms, the departments may have different names or be merged (e.g., sales and marketing), but essentially these are the building blocks of a book publishing firm.

When you have submitted your final manuscript to your publisher, there is still more work to do during the production period. You may be asked for input into page layout and cover design. You will certainly have to answer queries from the copy editor and read through proofs of the book before it goes to print. The publisher may employ a professional proofreader, but it is still worth checking for yourself, for two main reasons. First, even professionals don't spot everything, and second, it's your name on the book so really the buck stops with you. If it is difficult for you to read proofs, whether due to lack of time or dyslexia or for some other reason, try to find someone capable who can check them for you.

Your publisher's production team should give you a schedule for this work at the start, so you can negotiate on any points that are inconvenient for you, e.g., if they want you to read proofs during a scheduled family holiday. They should be able to offer flexibility at the start of the process, though they will then expect you to meet your commitments.

You may also have to create or source an index or have the cost of an index deducted from your royalties by a publisher. It is our view that it should be a publisher's responsibility to commission and pay for an index for any academic or other non-fiction book. A good index is vital if a book is to be truly usable. Anyone can pull together an index in Word, but it's not likely to be much use to real readers. Creating a good index is a job for a trained professional, which requires payment. Authors who write books in their own time should not be forced to choose between creating their own inadequate index and paying someone else to make a proper one.

## Marketing

Marketing begins before publication and continues long afterwards. Your publisher's marketing department should work to promote your book, particularly around the time of publication. For example, they should:

- Post information about your book online well ahead of its publication date
- Market your book to relevant retailers, such as bookshops, online retailers, wholesalers, and academic or public libraries
- Include your book in their catalogue and on their flyers for specific events such as conferences in your field
- Send out review copies, including to people you find who are willing to write reviews or can otherwise promote the book to a significant number of people, e.g., by writing about it on a widely read blog
- Take your book to conferences, display it along with other books on their stand, and offer a conference discount

- Promote your book via their e-newsletter and social media channels
- Give you a jpeg of the cover for your own use
- Make flyers, at your request, for you to take to conferences and seminars

These steps will help but keep in mind that your publisher has a lot of other books to try to sell as well as yours. Also, some publishers do very little to market the books they produce, so you will increase sales by doing some marketing work yourself. If nothing else, you should link to the book in your email signoff, use the cover image on your PowerPoint slides, and ensure that the book is mentioned on your professional web page(s). If you use social media, you can also highlight it there. Some people have their book details printed on their business cards. A launch event can help with publicity, and your publisher's marketing department may be able to help you organise such an event.

If you want to do more marketing, here are some further suggestions:

- Send information about the book to any e-lists you subscribe to
- Send information about the book to your professional association(s) to include in their e-newsletter

- As a recent alumnus or alumna, contact your institution to see whether it will help promote your book
- Ask your current or former college or university, professional association or group, or employer for help publicising your book through their websites, newsletters, and other publicity channels
- Write one or more posts featuring the book for blogs with big readerships in your field, and publicise the blog post(s) at and after publication through social media
- Create a video about the book or some aspect of the book, upload to YouTube or Vimeo, and publicise through social media
- Create a podcast about the book or some aspect of the book, upload and publicise through social media
- Publicise the book itself through social media – don't keep saying 'buy my book', but promote any good reviews or positive comments you receive
- Write an article for the mainstream media based on, or featuring, your book

Although this list is not exhaustive, you are very unlikely to be able to do everything suggested above because of time and skill constraints. You may find it helpful to develop and use a marketing strategy. This sounds grand, but it is merely a brief document setting out what you plan to do each week or month. Marketing activities that you have taken a little time to think about and plan are likely to be more effective than ad hoc activities.

Working to promote and market your book is not essential unless you are contractually obliged to do so. If all you want is a new line on your curriculum vitae, marketing is probably not worth the bother. But if you want your work to reach as many people as possible, it's worth developing and using a marketing strategy for your book.

## PIPs

Although journal articles are most important for **Kris** at this early-career stage, she would also like to write a book and edit a collection. She knows that books can help to establish someone's reputation as an expert, and she is particularly attracted to the idea of writing a short book to get her findings out fast without having to write a full-length book. Kris is also considering ways to expand on her methodological work by including other scholars' contributions to an edited book. She would also like to collaborate on writing a textbook that adapts her innovative research method to a how-to guide for future researchers. Kris will include preliminary research on these options in her publication strategy.

**Ella** sees ways to adapt her writing for parents of LGBTQIA+ teenagers. She will extract sections from her thesis and other academic papers that are relevant to these parents, and expand on them. She is not particularly interested in writing or editing a book on her own, but as soon as she has her doctoral qualification, Ella is planning to approach more experienced authors with a view to co-editing a book. She has already defined a theme, conducted a competitor analysis, and outlined her own chapter, as well as made a list of potential contributors. Her intention is to share this work with a more experienced author and offer to do most of the legwork in return for mentoring in the co-editing relationship.

**Nathan** is very keen to write a book as he sees this as a great way to get his message out to a mainstream audience. He wants to apply scholarly writings into a practical how-to guidebook. This will mean condensing research-intensive parts of doctoral writings and expanding on the implications of the study. He contacted a publisher who is interested in his work. Nathan has written a proposal and a draft chapter and is hoping that the peer reviews will be positive.

## Chapter summary

For many people, seeing their name on the cover of a print book is a personal goal regardless of whether or not it supports a professional purpose. However, writing a book is an enormous undertaking, so the commitment is a serious one. The effort needed to write an academic book or textbook typically extends beyond the actual manuscript to include pedagogical and visual features. You might choose to share the work with co-authors or co-editors, in which case it is critical to find others who are reliable collaborative partners, and to develop a clear plan for completion of the project.

## Exercises and questions for reflection: planning and writing a book

Use templates and worksheets found on the book website or www.path2publishing.com to complete this chapter's exercises and learning activities.

### *Reflection*

Reflect on your life and career aspirations in relation to writing and publishing a book. What could you gain from publishing a book? What might you lose?

Are your skills and interests more aligned with writing a book, co-writing a book, or editing a book? What would potential readers gain from the experience of reading a book versus another form of publication? Who would benefit from reading about your research process or findings? Who would benefit from reading about your research experiences or insights?

**Exercises**

*Part 1: research your options*

1 What is there in your identified assets that you could extract, condense, expand, adapt, or apply to create a short or full-length book? Write down your answers.
2 Think of a publisher who operates in your field or, if you can't bring one to mind, check the spines of books on your shelves or in the relevant section of the library to find the most common publisher. Go to that publisher's website and find its information for authors and proposal form. Read those to find out what would be involved if you wanted to write a book for that publisher.

*Part 2: add to your publication strategy*

Write out goals and key steps to include in the publication strategy you started to outline in Chapter 2.

## Good practice points

- Take care to follow precisely the book proposal format and instructions provided by your chosen publisher.
- Be clear about your target audience(s) and their needs.
- Plan and manage your time well; meet your contractual deadlines.
- Be prepared to do some marketing if you want your book to reach as many readers as possible.

## References

Davis, M. A., & Blossey, B. (2011). Edited books: The good, the bad, and the ugly. *The Bulletin of the Ecological Society of America*, *92*(3), 247–250. doi: 10.1890/0012-9623-92.3.247.

Lemon, N., & Salmons, J. (2020). *Collaborate to succeed in higher education and beyond: A practical guide for doctoral students and early career researchers*. London, England: Routledge.

Nederman, C. J. (2005). Herding cats: The view from the volume and series editor. *Journal of Scholarly Publishing, 36*(4), 221–228.

Salmons, J. (2017a). Envisioning an edited book. Retrieved from https://www.methodspace.com/envisioning-edited-book/

Salmons, J. (2017b). Reimagining ancillary materials for texts and academic books. Retrieved from https://www.methodspace.com/envisioning-edited-book/

Thomas, R. M., & Brubaker, D. L. (2008). *Theses and dissertations: A guide to planning, research, and writing* (2nd ed.). Thousand Oaks, CA: Corwin.

# 5 Why, when, and how should I publish book chapters?

**After studying this chapter, you will be able to:**

- Identify the value for contributed chapters.
- Understand the steps for developing and publishing a chapter.
- Analyse ways to use your thesis, dissertation, or other academic writings as the basis for book chapters.
- Evaluate implications for your publication strategy.

## Overview

Edited books are usually collections of chapters by different authors. Some may also include – or even be made up of – other forms of writing such as poems, stories, case studies, or comics. These books may be scholarly, professional, or practical. The editor or editors typically produce a proposal for a publisher, outlining the concept and explaining the value of the book. We looked at the opportunities and challenges of editing a book in Chapter 4; in this chapter we will focus on the role of the contributor.

A good edited collection can be an incredibly useful resource, bringing together a range of perspectives on the same topic in one publication. Also, edited collections can be a good place to publish important and useful work that doesn't get much of a chance in most academic journals. For example, in the field of research methods, tales of when methods don't work, or of ethical difficulties related to methods, can be hard to find in journals. Yet these kinds of narratives can be really helpful for learning and reflection.

There are a number of reasons you might decide to write a chapter for an edited collection. These could include:

- You have something to say and the proposed book offers an appropriate platform.
- You want to add to your curriculum vitae (CV).
- Book chapters are often shorter than journal articles.
- Book editors may give encouragement, helpful feedback, and even coaching to inexperienced writers.
- The other authors are scholars with whom you would like to be associated.

- The collection focuses on an important or emerging theme, and you want to be recognised as a thinker on the topic.
- You would like a copy of an otherwise expensive book.
- One of the editors is a friend, or someone to whom you owe a favour.
- You'd like to try writing a book chapter.

There may also be reasons against writing a book chapter. These could include:

- You are short on time.
- You are unlikely to receive royalties, though you may receive a small honorarium or a gift card for books from the publisher.
- The word limit is too constraining.
- The other authors are scholars with whom you would not like to be associated.
- Publication elsewhere, for example, in an academic journal, would be better for your career.
- Publication elsewhere, for example, self-publishing, will get your work out faster.

Of course, these lists are not exhaustive. They are designed to start you thinking about the pros and cons of writing book chapters, and to help you work out whether there are other reasons for or against in your particular case.

## How to use your identified assets in book chapters

Use your identified assets to create book chapters (Figure 5.1) with one or more of these approaches:

- When you look at your assets, are there sections that could stand alone? If so, could you *extract* a section and use it as the basis for a book chapter? A clear description of a research problem, literature review, ethical dilemma, and/or methods chapter might be readily presented as a chapter.
- You might need to draw on elements from the entire thesis or dissertation, and *condense* them to fit the length and format of a book chapter.
- To *expand* on your assets, add elements that will strengthen and clarify the original writing. Adding specific examples could be beneficial.
- You could find that the requirements of the edited book are different from the academic focus of the dissertation or thesis, so some *adaptation* may be crucial.
- The edited book may be intended for a general audience or an audience of practitioners. These readers are less interested in theory and scholarly foundations, and more interested in practical steps they can use to improve their lives, work, or organisations. Therefore, your writing will center on how to *apply* findings and insights.

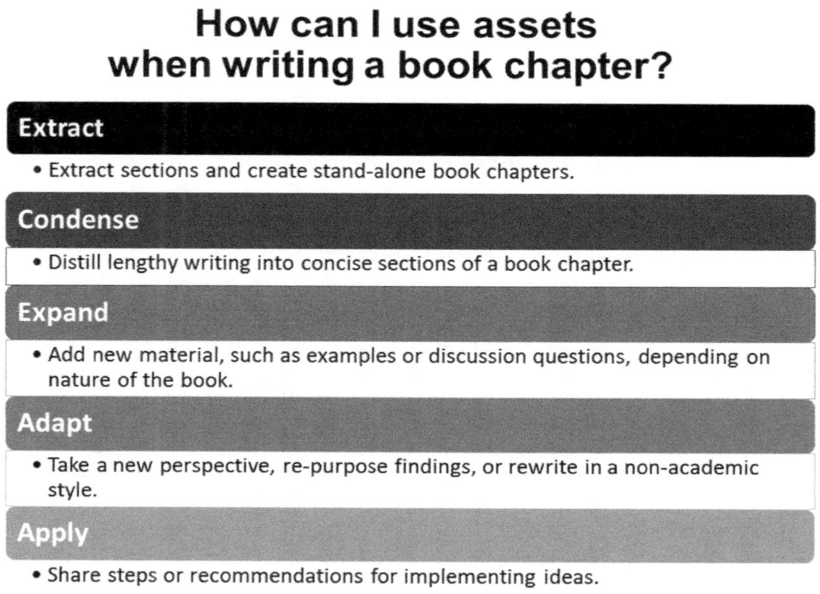

# How can I use assets when writing a book chapter?

**Extract**
- Extract sections and create stand-alone book chapters.

**Condense**
- Distill lengthy writing into concise sections of a book chapter.

**Expand**
- Add new material, such as examples or discussion questions, depending on nature of the book.

**Adapt**
- Take a new perspective, re-purpose findings, or rewrite in a non-academic style.

**Apply**
- Share steps or recommendations for implementing ideas.

*Figure 5.1* How can I use assets when writing a book chapter?

## When an edited collection finds you

A collection may be edited by someone who finds you because they know you well or distantly. Perhaps they are or were your doctoral supervisor, or maybe they heard you speak at a conference. It could be that they saw your abstract in a conference programme, or found you online, or through social media. Whatever the route, they know – or know of – you and your work, and think a chapter from you could be a good fit for their collection.

A direct approach from a book editor can feel very flattering, and indeed it may be so, but we would suggest you don't take that at face value. It may also be the case that the editor is desperate to fill their pages. Some edited collections are poor quality, slapped together with little thought and a lot of exploitation, badly edited and with no index. You have every right to question the editor about their plans for the book, what they want from you, and why. In fact, we would suggest you do so, politely and professionally, of course, because that can help you decide whether or not to write the chapter. You might want to find out who else will be writing chapters and how diverse the authors are. It is also worth asking about the publisher that the editor has chosen to work with: Why did they choose that publisher?

What commitments has the publisher made for production values and distribution/marketing of the book?

If you decide to go ahead and write the chapter, you will do a better job if you're aware of the editor's thinking. We would advise you to ask for a copy of the proposal the editor submitted, or plans to submit, to the publisher. The proposal will show you where your chapter fits into the book, and also what else is included so you can cross-refer as necessary.

## When you want to find an edited collection

Another option, when it comes to book chapters, is to decide you want to write one and go looking for a suitable edited collection that is currently under construction. Editors often create a call for chapters that spells out the topics, writing styles, chapter length, and formats they want to include in the book. The call for chapters should also include details of timelines and checkpoints for stages of the publication process. Calls for chapters may be distributed on the publisher's website and/or by editors through email lists and social media networks.

You will need to submit a proposal for the chapter you plan to write. Make sure you follow the requirements and time frames set out in the call for chapters, and provide the editor with any details they have requested. This helps to demonstrate that you would be a valuable contributor, and makes it easy for the editors to determine whether or not your proposed chapter would fit.

The editor may decide on which chapters to incorporate into the book, or an editorial board may have a role in the decision. For scholarly or academic books, a peer-review process would be used to determine which authors are invited to develop their proposal into a chapter. Even in these cases, the editor will probably check the proposal to determine whether it meets the minimum requirements of the book before sending it out for peer review.

As well as receiving an invitation or finding a call for chapters, you could also decide to publish a chapter in a book you edited yourself, whether alone or with others. There is more information in Chapter 4 on how to edit your own collection.

## How to plan a book chapter

The editors should have given you a word count, which will probably be in the region of 4,000–6,000 words. This is likely to be shorter than any chapter in your thesis or dissertation (and also shorter than most academic journal articles) so you need to plan for succinct writing around one specific theme. Bear in mind that a book chapter needs to contain original material. It's fine to use material that was original in your thesis or dissertation, but remember that is most likely to come from your findings, conclusion, and perhaps your methodology chapters. It is less likely to come from your literature review and analysis chapters. Also, there may be a good basis for a book chapter among the as-yet-unpublished assets you have identified.

Map out the structure of your chapter, devising headings to guide you. Under each heading, give a brief description of the intended content and an approximate word count for that section. Give your chapter a working title. Sketch out the literature you want to cite and what you need to do (if anything) to update your reading.

## How to write a book chapter

When you have your draft structure and working title, open your identified assets and import the relevant sentences, paragraphs, or sections into the draft structure. It will look messy; don't let that worry you. This exercise can take some time and is worth doing carefully to make sure you have imported all the relevant material. Once you're done, take a look at your embryonic book chapter. How many words do you have? Are there any headings with no material? Is there any repetition? This will help you figure out what you still need to do.

What you still need to do comes under five headings:

1 Reading
2 Rewriting
3 Writing
4 Editing
5 Polishing

Listing them in this way can make them look as though they are discrete and consecutive processes. That is not the case. For example, if you have twice as many words as you need, it is best to start with some editing. Not everyone has to do reading, and those who do may not need to do much. But separating out the different aspects of the task in this way helps you to think about and plan the rest of your work.

It's fine to change the structure as you develop your ideas – in fact, it's almost inevitable – and the title may change too. However, the word count is probably not negotiable, though if you really need, say, an extra 500 words, you could always appeal to the editor.

## Working with the editor(s)

Writing book chapters can feel and be quite a lot more collaborative than writing academic journal articles, even if you're a sole author. Editors are usually invested in the quality of the book, and may be more friendly and communicative than journal editors. They may even offer a form of coaching or mentoring for inexperienced contributors. Also, you are unlikely to have to tangle with unwieldy online submission systems. You may be asked to serve as a peer reviewer for one or more other chapters of the book, or to help in another way, such as by contributing to the glossary or appendices.

It is a good idea to stay in touch with the editor about anything to do with the process and practicalities of producing your chapter. If you think you might want to edit a book yourself one day, it is worth observing the way your editor operates, and making brief notes on what you would like to emulate, anything you would do differently, and why. There is more information on editing collections in Chapter 4.

## Seeking and working with pre-submission feedback

Once you have a draft chapter, it is often a good idea to seek feedback informally before submitting your chapter to the editor. Think about whether you have a friend or colleague whom you could ask to read your draft with a constructively critical eye. It is not so crucial to do this for a book chapter as it is for a journal article, but it is still a step worth taking, particularly if you are an inexperienced writer. And do use the feedback you receive to revise your work before you send it in to the editor. There is more about how to find and work with feedback in Chapter 3; so if you have come straight to this chapter, we would advise you to read the sections on pages 73 and 74 of that chapter.

## Dealing with reviewers' comments

There are three main types of reviews for book chapters:

1 Standard anonymous peer review, which could be by anyone
2 Anonymous review by another chapter author
3 Review by a named chapter author

Advice on working with standard anonymous reviews can be found in Chapter 3. Responding to reviews from other chapter authors is a little different. If they are named, it's probably easier because you can have a dialogue with them about anything that is unclear or where your views diverge. On the other hand, it won't be easier if you find them difficult to work with. If the reviewer is not named, you may still be able to work out who they are (and the temptation to try would surely be hard to resist). However, you need to stay within the bounds of convention, even if you're sure who the reviewer is, and treat their review as if it were a standard anonymous review. If anything, you should be even more polite and diplomatic, because being published in the same book as someone is a signal that you're on an equivalent intellectual level. Also, it gives you an opening if you ever want to contact one of your co-authors for a different reason. If they happen to be the person who reviewed your chapter, and they don't like the way you acted on their review, you risk a cool reception.

## Reviewing others' work

The implications of this, of course, are that you too may be asked to review one or more chapters by your fellow authors. Reviewing can be quite a challenge at first. Most of us have to battle with self-doubt, worrying about whether we'll have anything useful to contribute. Put all that aside and read the draft chapter carefully, making notes of anything that occurs to you. Then use these questions as a guide to your review:

- How well does the chapter conform to the word count and any other guidelines?
- Is it of suitable quality? (See Chapter 1 for ways to assess the quality of written work.)
- Does it fit with the book as a whole?

Write your review politely and professionally. Take care to praise the good points as well as identifying where there is room for improvement – and, crucially, how that improvement could be made.

If you think the chapter you're reviewing is really poor, you should be able to raise this with the editor for advice about how to handle the situation. The editor won't necessarily read draft chapters before sending them out for review so they should be glad of your input. Also, they will have some kind of relationship with each chapter author, which should enable them to help you resolve any problems.

## PIPs

**Kris**, on their way to a career as a professional academic, may not need to write book chapters depending on their field as well as their institution.

Some fields regard books and chapters more highly than others, and some institutions recognise chapters as viable contributions. Kris will need to ask more experienced people or look at tenure and promotion guidelines before deciding. They might decide to contribute a chapter if there is something they could write easily, perhaps updating previous writing, or if they want to see their work published together with leading thinkers in their field.

**Ella**, our independent scholar, thinks a book chapter could be a useful advert for her ability to conduct and write about research. The lack of payment puts her off, but another reason to contribute could be if she wants a copy of the book and it won't be affordable otherwise. So she decides to look favourably on direct invitations to write book chapters, and to respond to calls for chapters by known and/or respected scholars. However, she does not plan to seek out calls for chapters by unfamiliar editors.

**Nathan** would write as many book chapters as he could, because he is so keen to make his findings available to as many people in as many ways as possible. He plans to adapt his findings for readers who have family members on overseas military service and for readers who are mental health professionals working with such families. The only thing that would deter Nathan is lack of time, as he is mindful of his own mental health so tries not to take on more than he can manage. He searched the internet and was delighted to find calls for chapters for three relevant books. However, he knew he wouldn't have time to write all three, so he chose the one that seemed likely to have the widest distribution as it is with a reputable international publisher.

## Chapter summary

Chapters contributed to edited collections can be a good part of a larger publication strategy. However, much depends on the nature of the book and the quality of the editor or editorial team. At their best, book editors have commitment to creating an excellent volume of chapters that represent the perspectives of diverse contributors. Their commitment to the book can translate into a more personal and hands-on approach than is typically found in academic journals. New writers can gain credibility by having their work appear side-by-side with experienced and respected scholars.

## Exercises and questions for discussion and reflection: planning and writing a book chapter

Use templates and worksheets found on the book website or www.path2publishing.com to complete this chapter's exercises and learning activities.

### Reflection

What edited collections have you read? What did you like or dislike about them? Why?

### Exercises

*Exercise 1: research options*

1 Identify content from your thesis or dissertation that could be extracted, condensed, or adapted to meet requirements for book chapters.
2 Identify publishers' websites and/or email lists from organisations or professional associations and/or social media or other web pages where calls for chapters are distributed. Evaluate at least one call for chapters to help you understand the expectations and steps of the process.

*Exercise 2: add chapters to your publication strategy*

Write out goals and key steps and update the publication strategy you started to outline in Chapter 2.

## Good practice points

- Find out about the rationale for the whole book and its proposed contents.
- Check out the publisher and make sure you're happy with the way they work.
- Keep to the word count and follow any other guidelines.
- Write your chapter to fit within the whole concept.
- Be collegial and respectful.

# 6 Why, when, and how should I publish case studies?

**After studying this chapter, you will be able to:**

- Distinguish key types of published case studies.
- Understand how case studies are used.
- Draw material from existing writing suitable for a case study.
- Evaluate ways case studies could be beneficial to your publication strategy.

## Overview

This chapter offers an explanation of case types and options for publishing case studies. Case studies are sometimes overlooked as a publication option. However, they offer a unique way to get your work published, including aspects of your work that may be difficult to get published elsewhere. Importantly, case studies allow your work to be analysed, discussed, and used in practical ways. Cases are used in instruction, in both academic teaching and professional skills training. They are also essential planning, decision-making, and policy-making tools for the public and private sectors.

There are several kinds of case studies. We will look at instructional, research methods, and exemplary cases. Cases can be presented in text or multimedia formats. They can be published in a stand-alone form or as part of the materials included with training courses, or tutorials, books, or articles. After reading this chapter, you can think about whether writing cases aligns with your goals, and decide whether to include this option in your publication strategy.

## What are *case studies* and why would I publish one?

The term *case study* has a variety of definitions and applications. Qualitative and mixed methods researchers use case study methodologies to design and conduct research, and they may call the presentation of their findings a case study (Yin, 2014). Published cases bear some similarities to these research products in that they are focused and bounded by clearly stated parameters. The main difference is in intention: Published cases are intended to be analysed and discussed. Unlike other kinds of writing, the key issue or dilemma is not necessarily resolved. Readers are expected to take an active role and find their own answers or solutions to the problems being presented.

At its simplest, a case study is a particular kind of story. A case encapsulates an issue or dilemma, providing a snapshot of a situation or problem at a specific point in time. The reader learns about the problem by viewing it from the author's experience and/or other perspectives. By going through the process of identifying and analysing the issues, conducting further investigation of the problem, and proposing new solutions, insights can be gained about the problems and contexts.

### *Why should I add cases to my publication strategy?*

There are some real advantages to consider when thinking about whether writing case studies is right for you. Case studies are often quite straightforward to write as publishers usually provide clear guidance, and the text is likely to be based on events and literature with which you are already familiar. If impact is important to you, case studies offer a practical way that others can use your research. Readers can learn from your experiences, successes, and failures. By offering some element(s) of your study for analysis, you can further progress in your field by motivating others to conduct their own empirical research, or contribute to changes in practice by encouraging others to generate new solutions to complex problems.

### How are cases published?

Cases can stand alone or be created as supplementary materials to complement a book or an article. You can write a case using text alone, or create videos, games, or other interactive experiences that bring the case to life.

There are a number of ways to publish cases, depending on the audience you want to reach. Most major academic and business publishers offer case collections. You could also self-publish cases. For example, you could develop an active problem-solving case that instructors might buy to use in professional training workshops or in the classroom. (See Chapter 9 for more on self-publishing.)

If you are interested in writing a case study, learn about the types that exist, and look at cases that are used in your field or profession.

## Types of case studies

Three types of cases are most common: instructional cases, research cases, and exemplary cases.

- **Instructional Cases:** Have you had a learning experience centred on a case? Instructional cases highlight particular problems, leaving conclusions to the student. Instructional cases invite the students to learn about the problem and review options for solving it. They ask: 'How would you handle this situation?' Cases are often used in multidisciplinary collaborative and team projects. This kind of active learning allows students to gain insights and problem-solving skills not possible to acquire by simply reading about a problem and the ways others have addressed it.

Instructional cases are often distributed by textbook publishers and university presses. They are sold as single publications or as supplementary materials to textbooks, tutorial, workbooks, or educational media.

- **Research Methods Cases:** When you studied research methods, did you learn from the trials and errors of others? Methods cases offer readers a way to understand what really happens when the researcher conducts an interview or digs through an archive to find the right document. Research cases are narrative accounts of the conduct of an actual study from the perspective of the researcher (Salmons, 2014b). These cases show alternatives and options the researcher has considered, a rationale for choices made, and descriptions of how the design was implemented. A good research case also reveals the researcher's decision-making processes – including in-the-moment decisions about how to address problems that emerge when the inquiry does not go as planned. By exploring methods cases, student or novice researchers learn how the elements of a research design come together.

Research methods cases are primarily distributed by publishers of textbooks and resources for researchers. They are also distributed by research, writing, or dissertation/thesis coaches.

- **Exemplary Cases:** Businesses in marketing firms, government agencies, or non-governmental organisations sometimes need to represent the characteristics of a typical customer, participant, audience, or event. Organisations may choose to create exemplary cases when dealing with sensitive matters because a fictional case allows them to protect the privacy of customers or participants. Unlike the instructional and research cases described above, these cases are not constructed in an open-ended manner – they tell a complete story. These cases are meant to show in a concise way how some particular characteristics manifest, to help decision-makers understand users' needs or preferences.

Exemplary cases tend to be proprietary and may be used privately, within the organisation, or distributed as a part of a public relations campaign. Sometimes detailed demographic descriptions of the research population could be used as the basis of an exemplary case. The PIPs in this book are one form of exemplary case. If you are considering a career that involves decision-making, writing an exemplary case will demonstrate your understanding of the people and problems you want to address.

An example might help you to distinguish between these types. Let's say your doctoral research was a qualitative study that explored ways social entrepreneurs develop services to meet local needs in under-served urban neighbourhoods. You could:

- Develop an *instructional case* to be used for teaching students about identifying and addressing needs in diverse urban neighbourhoods.

- Develop a *research case* to show how you overcame obstacles and conducted this study. Discuss how you gained access and located participants in urban neighbourhoods where outsiders are often greeted with suspicion.
- Develop an *exemplary case* to profile sensitive issues you discovered in urban neighbourhoods, such as unemployment, drug use, or undocumented immigrants.

As this example shows, you can potentially develop more than one case, or more than one type of case, from your doctoral research.

### How to use your identified assets in cases

Could you use one or more of these approaches to create a case?

- To *extract*, take out a section of your thesis, dissertation, or other scholarly writings that describes a dilemma or research experience that could serve as the basis for a case study.
- To *condense*, summarise background information. Unless it is a research case, you will most likely need to condense the literature review and methods sections. Alternatively, think about ways you could condense lengthy

# Example: Using assets when writing cases

**Extract**
- Extract decision points or dilemmas.

**Highlight two angles to a problem:** A research participant described a dilemma when starting a business serving other entrepreneurs that was supported with ad revenue. She decided to be transparent about earnings, which built credibility with the entrepreneurs but clients placing the ads were not pleased.

**Condense**
- Distill key elements.

**Eliminate non-essential info:** Participant provided lengthy descriptions: extraneous material summarized or removed for case. Condensed scholarly background into a few succinct points; literature otherwise not relevant for a case.

**Apply**
- Develop case that invites users to apply findings for practical purpose.

**Readers apply knowledge to solve the problem:** Position the story as open-ended: what should the entrepreneur do to build credibility with both stakeholder groups, entrepreneurs *and* clients?

*Figure 6.1* Using assets when writing cases.

descriptions of a problem you encountered when conducting the study, or complex situations you discovered in the field when collecting data, to create a research case.

- To *expand* on your assets, add elements that will help readers learn from the case. Look for ways you could expand on existing writing to construct an open-ended case that would invite students or other users to apply new understandings to solve the problem.
- Create a case that invites readers to *apply* what you have learned and solve problems.
- *Adapt* your design and methodology to fit the format and audience for a research case.

You can extract and adapt specific elements from your thesis or dissertation to use as the basis for a case study. For example, you might extract problem statements from an early chapter, methods and ethics descriptions from your methodology chapter, and results from a later chapter, and adapt them for a research-oriented case. Or, you could focus on situations you observed in the field for a problem-centred case.

Alternatively, you can expand on previous work by drawing on notes or memos that were not included in the completed thesis or dissertation. Reflect on dilemmas or dead-ends you encountered in the process of understanding your research problem or conducting the study.

Think about the stories you've told your friends about the challenges that emerged, the crises or near misses that could have sabotaged the whole study. These are not the stories you wrote about formally in your thesis or dissertation. What can others learn from your experiences, be they good or bad?

To translate your experiences into a case study, you may find that you need to adapt informal notes you made or draw on observations that you didn't record at all. You will need to expand those personal stories into an organised discussion that fits the format and requirements of the publisher. If the research experience was a painful or unsuccessful one, it will be important to step back from the emotions and create a clear description.

Because cases have a practical function, they can be an important part of your publication strategy. The impact of your research can be extended when others use your dilemmas or research experiences in an instructional or professional context. You can build name recognition and credibility for your work (and other publications) by developing cases for use in instructional or professional settings.

If you are a PhD student or graduate, research methods cases can demonstrate your prowess at designing, conducting, and writing about empirical inquiries. By developing an instructional case, you can show the value you will bring to a faculty position that includes working with colleagues to develop curricula. For those in fields related to policy- or decision-making, an exemplary case could display your insights into current populations or issues.

If you are a professional doctorate student or graduate, cases can demonstrate your ability to bridge theory and practice. You can show your potential as someone who can identify and describe critical problems from multiple viewpoints. If you are considering a role such as a consultant or facilitator, you can develop cases that can be used in workshops to train practitioners.

## Writing case studies

Case writers are storytellers. It is almost like writing in a mystery genre: You want to offer the readers clues, but make them work to discover 'who done it'. We are accustomed to showing results that illuminate the research question or hypothesis, but this kind of writing points to the unanswered questions, conflicts that look different depending on the vantage point, or fuzzy situations where a solution is not obvious.

Writing a case study starts when you select a problem, dilemma, situation, phenomenon, or experience. Whether used in the classroom or professional skills training, case analysis uses critical and creative thinking. The goal for the case writer is to provide a story that readers can use as the basis for analysing the issues, conducting further research, discussing varied perspectives, and proposing solutions (Salmons, 2017).

### Writing cases for teaching and learning

To write this type of case, you might extract descriptions of the research problem, or adapt and apply relevant assets from your findings. When writing an instructional case, the challenge is to write in a way that will encourage students to build higher-order critical and creative thinking skills. You want students to discern underlying factors or influences, and come up with solutions that are not simplistic or obvious. Case analysis assignments, projects, or team activities should ask students to dig below the surface and build awareness about complicated real-world scenarios.

Learning potential for case analysis can be described using the categories from Bloom's Taxonomy (Krathwohl, Bloom, & Masia, 1964). Looking at these progressively more complex options can help you think about the kinds of potential learning opportunities you might want to build into your case (Salmons, 2014a):

- *Knowledge:* Students are asked to use cases and related literature to define terms, concepts, and approaches associated with the problem.
- *Comprehension:* Students are expected to understand the problem described in the case and be able to use this understanding to explicate key dimensions or identify stakeholders.
- *Application:* Students are expected to apply ideas or theories from the class into a situation presented in the case study.
- *Analysis:* Students are expected to distinguish motives and assumptions that underlie the perspectives of various stakeholders, and determine the problems presented in the case study.
- *Synthesis:* Students are asked to draw on multiple sources to develop new approaches to resolving problems described in the case. Based on their analysis, they develop recommendations for each key stakeholder.
- *Evaluation:* Students are expected to evaluate the problems and contexts presented in the case, develop alternative approaches, and judge the value and effectiveness of

recommendations they have chosen. In class discussion or team projects, they may also evaluate their analysis and recommendations in comparison to those suggested by classmates.

## Writing cases about research methods

To write this type of case, you might expand and adapt relevant assets in your memos, notes, communications with your committee, or other records you kept when designing, conducting, and writing about your doctoral research. Research methods cases can be read by students in a course setting, so the above points about teaching and learning will apply. But research cases can also be read by novice or early career researchers, or experienced researchers who are contemplating a new approach. The most useful cases for any of these readers is honest and presents the unvarnished truth of your experience. Were you terrified by your first research interview? Did you arrive at the research site to discover that your contact person resigned the day before? You wouldn't discuss these horror stories in a research article, but the tactics you used to turn things around will be very helpful to the case study reader.

### *Writing cases to profile examples*

To write this type of case, you might extract descriptions of the research problem site, and/ or population. You could expand and adapt relevant assets from your literature review, field notes or memos, as well as the findings discussed in your dissertation or thesis.

Exemplary cases serve a different function than do instructional or research cases. They are meant to provide rich descriptions that can be used to inform readers or help them make decisions. For example, a case about working mothers with children under the age of five, could be written to help policy-makers decide what kinds of child care are needed in the community. Or, it could be written to help businesses decide what products and services these women are likely to buy. An exemplary case could profile a situation, such as unemployment, gun violence, organisational change, or access to medical care. What are the characteristics of the population you studied? What situations did you study, or what situations did you encounter in the course of your research? How could what you've learned help someone who is trying to make decisions about how to meet the needs and interests of this group of people, or to create policies or social programmes that will help address problematic situations?

## PIPs

**Kris** decided to include case studies in their publication strategy – to be written after they get at least one journal article accepted for publication. They would like to develop a series of research cases that could be adapted from their dissertation to help graduate students learn about advanced quantitative methods.

**Ella** would like to write instructional cases that could be used to help students in the education field prepare for future work with LGBTQIA+ children and their parents. Prospective cases will expand on her findings, and apply them in practical ways to help educators. When she evaluates potential publishers for the book she is planning to write, she will explore whether they include cases as part of a book package. If not, she will consider self-publishing cases that she can sell through her independent consulting entity.

**Nathan** does not want to publish stand-alone cases. Instead, he plans to integrate cases into the practical guidebook that is central to his publication strategy. These cases will be used to illustrate common problems faced when family members are on overseas military service. He will show counsellors and military family members reading this book how the findings generated through his research apply to real problem-solving.

## Chapter summary

This chapter introduced the idea of case writing as a valuable option to consider as part of your publication strategy. While the use varies by the type of case, this kind of writing creates research impact because instead of simply reading about your study, users dig deeply and look for innovative solutions to the problems themselves or the research approaches used to study them.

## Exercises and questions for reflection: planning and writing a case study

Use templates and worksheets found on the book website or www.path2publishing.com to complete this chapter's exercises and learning activities. Visit www.path2publishing.com to read sample instructional, research methods, or exemplary cases, and find resources about cases and potential publishers.

### *Reflection*

What types of case analysis did you participate in as a student? What do you wish the case writers had included in those cases?

**Exercises**

*Part 1: explore case study publication options*

**Exercise 1:** Case studies are more common in some fields than others. Explore ways case studies are used in your field or discipline. Categorise examples into the types described above: instructional, research, or exemplary.

**Exercise 2:** Who published the cases you found in Exercise 1? Look on the publishers' websites, and search for case development divisions of research centres or institutes. Find and evaluate publishers or other distributors where you could submit proposals for case studies. Use Table 6.1 to record what you find, and the specific steps that will be needed for each option.

**Exercise 3:** In addition to reviewing your audit from Chapter 1, look at some of your background materials. Identify content, including research notes; emails to supervisors, committee members,

*Table 6.1*  Case study requirements

| Publisher | Case Type | Proposal Requirements | Additional Information |
|---|---|---|---|
|  |  |  |  |
|  |  |  |  |

or fellow students; design notes; and proposals written before the study was conducted. Pay attention to problem statements, theoretical frameworks, literature reviews, method sections, insights, models, and/or findings that could be adapted to meet requirements for research. What stories could you tell that might help others to understand particular situations or generate new solutions?

## *Part 2: add cases to your publication strategy*

After you complete the exercises and reflect on publication options that align with your goals, update the publication strategy you started outlining in Chapter 2 with any case-writing options that interest you.

## Good practice points

- Consider assets that might not suit other publication types including notes or memos about obstacles you confronted.
- Think about the best use of your research for instructional, research, or exemplary cases.
- Identify grey areas or problems without obvious solutions that lend themselves to case

studies that could motivate students or other readers to consider different approaches to addressing the issues.

- Look for publishers of case studies in your professional field or academic discipline and subscribe to email lists where calls for cases are announced.

## References

Krathwohl, D., Bloom, B., & Masia, B. B. (1964). *Taxonomy of educational objectives: The classification of educational goals Book 2: Affective domain*. New York, NY: David McKay and Company.

Salmons, J. (2014a). *How to use cases in research methods teaching: An author and editor's view*. Retrieved from http://methods.sagepub.com/case/how-to-use-cases-in-research-methods-teaching-an-author-and-editors-view. doi: 10.4135/978144627305014534935.

Salmons, J. (2014b). *What are research methods cases and how might they be used?* Retrieved from London: http://nsmnss.blogspot.com/2014/02/new-social-media-new-social-science-and.html

Salmons, J. (2017). A case for teaching methods. Retrieved from https://www.methodspace.com/case-teaching-methods/

Yin, R. K. (2014). *Case study research: Design and methods* (5th ed.). Thousand Oaks, CA: Sage Publication.

# 7 Why, when, and how should I work with publishers?

**After studying this chapter, you will be able to:**

- Identify kinds of relationships and expectations involved while working with publishers.
- Understand how to plan your work when publishing with traditional book and journal publishers.
- Evaluate implications for your publication strategy.

## Overview

Authors need publishers if they want to distribute texts and other books through respected channels, and publishers would not exist without authors. We use the phrase 'commercial publishers' to denote publishers of books and academic journals. The world of publishing has changed a great deal in recent decades. At the turn of the century, all books and journals were created in hard copy and distributed through bricks-and-mortar booksellers and academic and public libraries. The internet has changed all that. As discussed in Chapter 4, now there are e-books, audiobooks, as well as hard copies of print books. Online services enable diverse formats for publications to be created by authors as well as – or instead of – publishers. These alternative publication options will be discussed in Chapter 8.

The internet also led to the open access movement of the 21st century. The aim of this movement is to make publicly funded research available to the public rather than being hidden behind paywalls and in expensive hardback books that are available only to academics. Much academic research and writing is now made openly available through institutional repositories, though these can be difficult to search. There are also an increasing number of open access journals and books. However, these are by no means universal, and publishers are struggling to reconcile the open access ideal with the real-world need to make at least enough money to pay staff and cover costs, and in some cases enough to create a surplus and pay shareholder dividends.

Given these seismic changes that are still sending waves through the industry, commercial publishers have to innovate to stay relevant.

Many are doing this effectively, yet much of their work is still conducted in traditional ways. And their work is still highly relevant to mainstream academia where textbook and academic publishers have long-standing relationships with academic libraries and bookstores.

## Business models of different types of publishers

Academic book and journal publishers are often one and the same. They may have a name you recognise, such as Palgrave Macmillan, Routledge, SAGE, or Harvard University Press. However, the name does not necessarily mean that the publisher is an independent entity, though it may have been at one time. For example, Palgrave Macmillan is now part of the global multinational Springer group of publishers. Routledge is part of the bigger publisher Taylor & Francis, which in turn is part of the global multinational company Informa, which has many interests other than academic publishing. SAGE is an independent for-profit international publisher, and Harvard University Press is a division of Harvard University.

The business model can matter because different types of publishers offer different qualities of author experience and product, and different levels of royalties. It is important to do your research to determine which publishers offer the characteristics you want in regard to production values, indexes, cover design, marketing, and work to distribute the books they publish.

## How journal publishers work

Every academic journal has its own editorial or advisory board, usually made up of senior academics with good knowledge of the journal's subject matter. Board members advise the editor, act as advocates for the journal, and oversee its development. Their professional prestige will be linked with the perceived prestige of the journal.

Roles of editorial or advisory board members may include:

- Advising on the strategic direction of the journal, taking into account all relevant contextual factors
- Giving feedback on published issues
- Identifying topics for themed special editions
- Providing content by writing editorials, articles, book reviews, etc
- Encouraging others to provide content by submitting articles, etc
- Finding peer reviewers and/or undertaking peer reviews
- Sharing journal articles and calls for special editions via email and social media

If you know an editorial or advisory board member for a journal that interests you, it can be very useful to talk to that person about your plans. They can give you a great deal of insight into the workings of the journal and their advice can be invaluable. Otherwise, your first point of contact is likely to be the editor or an automated online submission system.

Journal editors are often the first point of contact for aspiring article writers (Murray, 2009, p. 65). Contact details should be on the journal's website (though this is not always the case). It's fine to email them with queries, though they won't appreciate a stream of emails asking about specific details, particularly if those are already covered in the information for authors available online. Editors will appreciate evidence of existing knowledge of the journal, so do your homework first and only contact a journal editor if there is no other way to answer your question. One thing you might ask a journal editor is: 'I am considering submitting an article about [topic] to your journal; in principle, would this be of interest?' This should reveal whether they already have a similar article in the pipeline.

Some journals ask you to submit your article to the editor by email; others use online submission systems such as Manuscript Central or ScholarOne. These can be annoyingly complicated to use, especially for the first time. Another occasion to email a journal editor is if you find a discrepancy between the information the journal provides to authors and the requirements of the online submission system. Only the journal's editor will be able to resolve any such problems. For all these reasons it is a very bad idea to leave yourself short of time to submit a revised manuscript to meet an agreed deadline. Technology can always go wrong and online submission systems are no exception, so make sure you have plenty of time for contingencies.

## How to choose a journal publisher

There are hundreds of thousands of academic journals in the world and, even if you are only looking at journals in a particular field, it can be hard to decide which journal to approach with your work. Deciding factors may include such things as impact factor, whether the journal is open access (OA) for readers, article processing charges (APCs) levied by some OA journals from authors, circulation of hard copies, and online availability. This depends on your priorities. For example, Kris would be likely to choose a journal with a high impact factor, which would be most help with her academic ambitions. Nathan, however, would prefer an OA journal with a big circulation and good online availability, to help his findings reach as many people as possible.

Other factors you may want to consider include things such as word count and submission system. Once you decide on which criteria are relevant to you, it can be useful to make a shortlist of possible journals to help you compare them in more detail.

One of our PIPs, Ella, did her doctoral research in the United Kingdom and has decided to write an article about her findings around the relationship between the levels of teachers' support for LGBTQIA+ pupils and overall school success rates. As we saw in Chapter 4, Ella's supervisor has agreed to co-write, and has set Ella the initial task of finding a suitable journal. Ella has identified four potential journals and has compared them.

Ella has also discovered that the *British Educational Research Journal* is published by Wiley, while the others are published by Taylor & Francis, and that each of these journals takes submissions through ScholarOne. This is all useful information to help her decide where to try to submit her article.

As with any publisher, you will get the best results from working with traditional publishers if you are polite and professional. Many journals, particularly those with good reputations, get far more submissions than they can publish. To maximise your chances of success, follow author guidelines to the letter. See Chapter 3 for more information about writing and submitting journal articles.

## How to choose a book publisher

Think about what type of publisher fits the type of book you want to write, and reaches the people who will buy and read your book. Is there a publisher who is particularly well regarded in your field? If you want to write a textbook, is there a publisher that has distribution agreements with university bookstores? If you want to write a general audience book, is there a publisher that has distribution agreements with local bookstores and libraries? Do potential publishers offer books in the format your target readers prefer: hardcover, paperback, e-books, audiobooks? Would you rather be a big fish in a small pond or a small fish in a big pond? Do your personal ethics encourage you towards an independent or a non-profit publisher?

It is worth investigating any publisher you are considering working with. Find out as much as you can about their corporate structure, their ethos, the books they publish, and so on. See Figure 7.1 for questions to ask. You should be able to do this research online by looking through their website and searching for documents such as their annual report and accounts, which can

# Questions for conventional publishers

**Publisher?**
- Respected publisher(s) in your field?
- Acquisition & submission process?

**Review process?**
- Who decides whether the work is accepted?
- What are the steps?

**Timeline?**
- How long is the acceptance process?
- How long between submission and release?

**Promotion?**
- Will you be expected to promote the book, what are the roles of marketing staff?

**Royalties?**
- Will you be paid up-front, with sales, or at all?
- What agreement or contract will you sign?

*Figure 7.1* Questions for conventional publishers.

be very informative and helpful (though not always easy to find). Do you know any of their authors? If so, contact them and ask searching questions about what the publisher is like to work with. If not, you could try contacting an author of a book you like from that publisher, tell them what you like about their book, and ask them about their publishing experience.

There is more information to help you choose a book publisher in Chapter 4. The one kind of publisher you should never, ever work with is any publisher who asks you to pay to publish with them. This is known as 'vanity publishing' and there will be more information about this in Chapter 9 on self-publishing.

## Negotiating with book publishers

Publishing is a business deal, so negotiating is essential if you are to get the best for you and your book. Understanding how this process works will help you plan your tactics, and prepare to engage with the publisher from the contract through to the final delivery of the book.

Position the conversation as a business deal by saying something like, 'As we haven't done business together before …' Second, ask for more than you think they will agree to. There's a chance they might say yes, and if not you are leaving space for them to make a counter-offer lower than your request but still considerably better than their original offer.

The publisher is likely to specify separate royalty rates for hardbacks, paperbacks, and e-books. You can negotiate harder on e-book royalties, because publishers' costs for e-book production and handling are much lower than for print books. They don't have to pay for paper, printing, storage, shipping, or returns. However, in some countries e-books incur more tax than print books. Where this occurs, publishers may choose to absorb the cost rather than pass it on to customers, in which case it would be a factor in negotiations. Also, these days some publishers are using e-book royalties to offset other business costs – but nevertheless it is worth negotiating, because you never know what you can get until you try.

If you get stuck on a figure that doesn't seem enough to you but the publisher won't budge, you can ask for a 'riser'. That means after, say, 1,000 copies have been sold in that format, your royalty will go up by a few percentage points. This is often easier for publishers to say yes because if they sell 1,000 copies, they have already recouped most, all, or more of their investment in your work, so then they can certainly afford to pay a higher royalty. It's probably not worth asking for a riser for hardbacks, as they don't sell many copies, but it is well worth giving it a go for paperbacks, especially if you're writing a book that is likely to have a wide readership.

The US-based Textbook and Academic Authors Association conducted an anonymous survey of academic textbook authors in 2015. They found that average royalties for textbooks ranged from 9%–14% and the highest royalty reported was 30%. However, some academic publishers offer around 3%–10% and it can be lower for more niche research books. Clearly publishers need to earn money from the books they publish, to pay for their staff, buildings, and other costs associated with their business. But some publishers operate with more fairness to authors than others. An interesting comparison is that art galleries often give artists around 50% of any sales made through their exhibitions. No doubt the economics of publishing and art galleries are different, but the economics of writers and artists are probably quite similar. Inexperienced writers often assume that publishers are doing them a favour by publishing their book and, as we have seen, this is not the case. Commercial publishers would not exist without authors, while authors can choose to self-publish and distribute their work without publishers. (See Chapter 9 for more on self-publishing.) Another assumption often made by the uninitiated is that the royalty figures offered to authors by publishers are fixed. This is not always so. Don't be afraid to ask for a higher rate than is initially offered.

Bear in mind that the person you are negotiating with, usually your commissioning editor, will not have the power to make the final decision. Keep your relationship with them as cordial and professional as possible, and make your case as clearly and concisely as you can, because you need them to advocate for you within their organisation.

Occasionally an academic publisher will offer a small 'advance' of a few hundred pounds. This is not an extra advance payment, it is an advance on royalties, which the publisher will claw back from your royalties after publication until it has been fully repaid. If a few hundred pounds would make a real difference to your work for the book – enable you to buy other books, for example, or to travel for meetings or to interview people – then by all means accept. But be aware that it's not extra money, they're simply rearranging the offer.

In many ways this is the simple part of the negotiations. Once you agree on the royalty figures, the publisher will issue a draft contract. It is a really good idea to get independent professional advice on that contract, because it will be hard to understand its implications unless you have specific legal expertise. In the United Kingdom, you can join the *Society of Authors* as soon as you have a draft contract, and specialist vetting of that and any other contract you receive is included in your membership fee. They will tell you which points to negotiate on, and how. In the United States, the *Textbook and Academic Authors Association* offers guidance on book contracts, royalties, and negotiations. (See our website, www.path2publishing.com, for links to these and other services.) Some countries have equivalent organisations but do check whether they offer what you need before you join. Otherwise you can employ a solicitor, though that is likely to cost more, and again please make sure they have the necessary expertise before you engage their services.

## Working with book publishers

When you have signed a contract, your commissioning editor and other staff at your publisher become your colleagues. You are all part of 'Team [Your Book]'. As with all colleagues, most are easy to work with, and now and again you come across an annoying one or someone having an off day. As in all work contexts, stay polite and professional. Communicate when you need to; don't be a complainer; get your work done; meet your deadlines.

When dealing with anyone from a book publisher, it is useful to remember that although your book is hugely important to you, it is only one of dozens or hundreds that they have to deal with. Whatever you are trying to achieve, acting like a diva will not help. Respecting that staff members of publishing firms are busy professionals will do far more to advance your cause.

## PIPs

**Kris** is keen to work with commercial publishers because it is books and journal articles that will help Kris most with their ambitions to become a professional academic. In particular, publishing journal articles is highly relevant for them. Kris is planning three journal articles in the hope of maximising their chances of getting at least one accepted. They have extracted three elements of their dissertation: a report on their findings for a politics journal, another report on their findings for an ethnicity journal, and a report on their innovative approach to quantitative research for a methods journal. They plan to identify suitable journals and then write all three articles, taking care not to duplicate between the first two.

**Ella** struggles with the economics of commercial publishing where the writer receives little or no money for time they put in to writing articles. Ella is more interested in independent publishing through blogs or self-published e-books, though she does recognise that a conventionally published journal article or two will help her to establish credibility with other academics and institutions. Ella and her supervisor have decided to submit to the *British Educational Research Journal*. Her supervisor is seeking funding from within the university to pay the APC so that, if the article is accepted, it can be open access. Ella will be second author on the article but she doesn't mind that, it'll be a publication with her name on it, which is the main thing. Also, she figures her supervisor deserves first authorship for giving her so much help.

**Nathan** is interested in maximising his audiences, and will happily publish with any journal or book publisher who can offer his work a wide circulation. In Chapter 4, he had submitted a proposal for, and draft chapter of, a guidebook to a publisher and was waiting for peer reviews. Those reviews were positive and constructive and Nathan was happy because he could see that implementing most of the reviewers' suggestions would improve his work.

## Chapter summary

There are several types of commercial journal and book publishers, and understanding their different business models is helpful when you are evaluating which one you want to approach. University and non-profit publishers approach books and journals differently from business-oriented publishers. Prospective writers need to understand how publishers will use their intellectual property, so they can decide what parameters to set. Researching and learning about the business of publishing enables writers to negotiate based on mutual respect.

## Exercises and questions for reflection: planning to work with a publisher

Use templates and worksheets found on the book website or www.path2publishing.com to complete this chapter's exercises and learning activities.

### *Reflection*

How good are your negotiating skills?

Negotiating and working directly with a publisher can be intimidating. Reflect on any concerns about this process and interactions with an acquisitions editor. Think about what you can do to prepare, so you will feel more confident.

## *Exercises*

### *Part 1: research options*

1 Identify three publishers of books and journals that are active in your field. One should be a non-profit publisher, one an independent for-profit publisher, and one a for-profit publisher that is part of a larger organisation with shareholders. Try to find the annual report and accounts of each publisher. Compare these documents, the company websites, and what people are saying about each publisher on social media. Does one appeal to you more than the others? Why?

2 What do you think are the ethical issues that pertain to academic publishing? Write down your answers, and then visit the website of the *Committee on Publication Ethics*. Have a look at their core practices, resources, and cases, to familiarise yourself with the main ethical issues in publishing from research.

### *Part 2: add to your publication strategy*

Write out goals and key steps. Include them in the publication strategy you started outlining in Chapter 2.

*Part 3*

Find trusted colleagues or friends with whom you can discuss how you will handle the kind of situations outlined in this chapter, and practice role-playing conversations you might have with an acquisitions editor.

## Good practice points

- Always be polite and professional when working with commercial publishers.
- Don't accept the royalties offered for a book, negotiate for a better deal.
- Remember that commercial publishers need you more than you need them, and an offer of publication is not a favour but a business proposition.
- Research book and journal publishers carefully to ensure you and your work are a good fit with them.
- Always get a draft contract checked by a suitable professional before signing.

## Reference

Murray, R. (2009). *Writing for academic journals*. Maidenhead, England: Open University Press.

# 8 Why, when, and how should I use alternative methods of publishing?

**After studying this chapter, you will be able to:**

- Understand motivations and options for alternative methods of publishing.
- Define key steps in the alternative publishing process.
- Analyse ways to use your identified assets as the basis for visual, audio, and other forms of alternative publication.
- Evaluate implications for your publication strategy.

## Overview

The use of alternative methods of writing and publishing is increasing in academia and research. For example, the education researcher Nick Sousanis submitted his doctoral dissertation in the form of a graphic novel. He was awarded his doctorate in 2014 and his graphic book, *Unflattening*, was published as a book by Harvard University Press (Sousanis, 2015). This is a rare example of a thesis or dissertation being professionally published as a book without much if any amendment.

Other alternative methods of writing from your identified assets include zines, comics, web comics, and research-based fiction. This chapter will give examples of each method as used in post-doctoral contexts and explain what is involved in creating them. We will also briefly cover songs, animation, video, and film based on research findings.

Some people find all this very exciting and inspiring, while for others it can feel much too far from the comfort zone. We would recommend aiming for a balance between enthusiasm and cynicism. It is useful to have more publishing formats for people who can and want to take advantage of those options. Those who don't want to engage with them needn't do so. However, we encourage you to read this chapter even if you are sure you never want to do anything other than write conventional prose. These formats are going to become more popular, not less, and while that doesn't mean you have to use them, it is wise to keep yourself informed. Historically people in academia have had to argue the case for using these formats, but that is changing fast, and the time may come when you would find it helpful to have a rationale for *not* using one or more alternative formats. If you understand what these formats are and what they can do, you are better placed to formulate such a rationale in coherent and considered terms.

And if you are one of the enthusiasts – don't get carried away!

## How can I use assets for alternative publications?

Figure 8.1 shows the questions that can be asked for alternative options.

## Types of alternative publications

### *Zines*

Zine is short for 'magazine'. The original zines were fanzines, self-published in small print runs by fans of musicians or other performance artists, and handed out at gigs or shows. Now there are academic zines including edited collections such as *So Fi Zine*, which publishes sociological fiction bi-annually. There are also authored works such as the *Market Café Magazine,* which focuses on data visualisation. *So Fi Zine* is available online and is free to download. *Market Café*

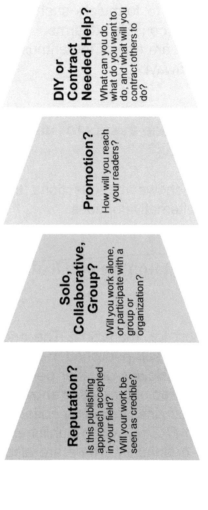

*Figure 8.1* Questions for alternative options.

*Magazine* is an independent publication, funded by its purchasers.

Zines were originally intended to be ephemeral, but some university libraries are now collecting and digitising zines. Also, some zines are now produced digitally such as *So Fi Zine.* Historically, zines were mostly in black and white, and quite small in size. These days there is no set format for a zine: It may be black and white or full colour or anything in between, mostly text or mostly images, long or short, large or small, created as hard copy or digitally, by hand or using software. So there are no rules, which is difficult for some people and liberating for others.

If you're going to create a zine, it is worth thinking about whom your zine is for and how you can give it to them. You could create a hard copy zine for an event such as a conference or seminar, or a digital zine for a community that you can reach online. Zines are a form of self-publishing so if you're interested in zines it is probably worth reading Chapter 9 of this book.

### Comics

Comics are short printed publications using sequential art, a form of communication with a very long history going back to Greek friezes and Egyptian hieroglyphs and maybe even some cave paintings. Comics can be drawn by hand or using software such as Comic Life or Pixton. If you're interested in using software to create comics, check online for the latest packages including open source options.

Comics are used in academia for teaching as well as for disseminating information (Duncan, Taylor, & Stoddard, 2016, p. 42). For example, one of the authors of this book, Helen Kara, has written a comic to support the teaching of qualitative interviewing. It is called *Conversation with A Purpose* and is available free to download from Helen's website.

If you have material in your identified assets that could be converted into a teaching resource, you might want to consider creating a comic. Writing for a comic is different because you need to imagine the visuals as well as telling a story. There is no set way to write a comic, but using a three-column table for your script can be useful, as shown in Table 8.1.

The left-hand column keeps track of the comic's format. Most of the numbers refer to a panel rather than a page. This comic is designed with a maximum of six panels to the page, though there may be fewer on some pages – the first page is a full-page chapter title image. The one-third page panel would form a strip across the top one-third of the page, the other four panels would each occupy one-sixth

*Table 8.1* Comic-writing plans

| Panel Number and Size | Text | Visuals or Storyboard |
|---|---|---|
| **Page 1** | | |
| 1 | Chapter 1 – Summer | Full page with big sun |
| **Page 2** | | |
| 2 – 1/3 page | | Summertime scene in park with children playing. Ali and Kaz visible under tree with picnic but not focus of scene. |
| 3 – 1/6 page | | Close-up on Ali and Kaz on picnic blanket, happy expressions. |
| 4 – 1/6 page | Kaz: I love picnics. | Close-up on Kaz's face looking even happier. |
| 5 – 1/6 page | Ali: Aarrgghh! | Ali leaping about slapping at a visible insect, picnic food flying everywhere. |
| 6 – 1/6 page | Kaz: What are you doing? | Kaz, on her feet, hands on hips, angry. |

of the page. The layout for page 2 of this comic would look like Figure 8.2.

If drawing isn't your forte, you can use software as mentioned above or collaborate with an

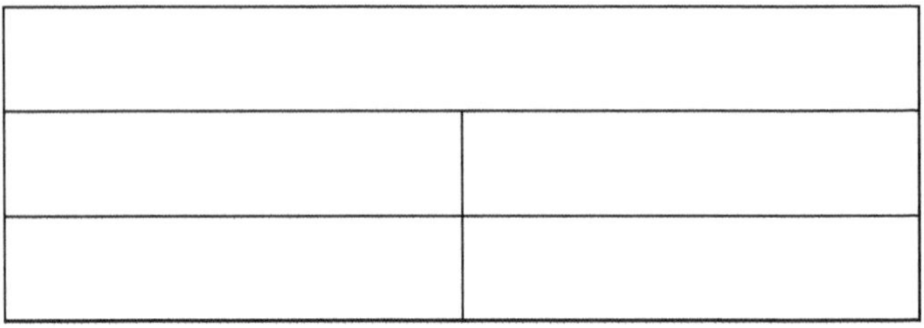

*Figure 8.2* Comic layout.

artist, as Helen Kara did for her comic. Unless you have a very obliging artist friend, you will need to pay for the artwork, and printing also costs money. Therefore, it would be worth applying for funding to support your project, especially if you want the comic to be free for readers. Another option to keep costs down is to self-publish through a platform such as Lulu or CreateSpace, so that people who buy your comic pay the printing costs for each one they buy. This was the route taken by Morris and colleagues when they published *Somewhere Nowhere: Lives Without Homes*, the findings of their research into homelessness (Morris et al., 2012). They interviewed around 100 homeless people in Stoke-on-Trent, and created five stories in comic form, which they published together in one book. This was partly to help disseminate their findings to participants, and also to make them publicly available to anyone who wanted to buy the book from Lulu.

## Web comics

These are very short comics published online. Each may have only three or four panels. Academic examples include *XKCD* and *PhD Comics*. Web comics are fun to create but, like blogs, they are time-consuming and content-heavy. (See Chapter 11 for more on blogs.) It may seem easy to think of a three- or four-panel comic. However, if you're going to create a worthwhile web comic, you need to come up with a three- or four-panel comic at least once a week, and that will be more challenging. On the plus side, if you are excited by the prospect of creating a web comic and have a good idea for such a venture as well as having (or being willing to develop) the necessary skills, it could be a great way to build an audience for your work.

## Graphic books

The line between comics and graphic novels or books is thin and blurred. Some would call the work of Morris and colleagues, mentioned above, a graphic novel. Generally, graphic novels or books are longer single works. Graphic books available today cover all sorts of subjects related to academia such as history, politics, society, culture, health, gender, and so on. Here are some examples, and you can find links to more on our website, www.path2publishing.com:

*Maus* by Art Spiegelman – Jews and Nazis in the 20th century

*Persepolis* by Marjane Satrapi – growing up as a young Iranian in the 20th century

*Palestine* by Joe Sacco – reports of the first Intifada in 1991–1992 in the West Bank and Gaza Strip

*A Thousand Coloured Castles* by Gareth Brookes – disability

*Naming Monsters* by Hannah Eaton – bereavement and grief

*Marbles* by Ellen Forney – bipolar disorder

*Blue is the Warmest Color* by Julie Maroh – young lesbian life

*Dykes To Watch Out For* by Alison Bechdel – adult lesbian life

*Special Exits* by Joyce Farmer – ageing and end of life

*Understanding Comics* and *Making Comics* by Scott McCloud – graphic books about comics

Of course, there are many, many more; this list is simply designed to give you an indication of the breadth of the genre.

There are also graphic books that don't use sequential art as such, but do use a lot of cartoon illustrations to help get their message across. These have great potential for communicating complex topics. One good example is *Queer: A Graphic History* by Meg-John Barker and Jules Scheele (Barker & Scheele, 2016). This bestseller is published by Icon Books who also publish the Graphic Guides series of short books on topics relating to science, English language and literature, philosophy and ethics, and psychology. Topics covered include feminism, capitalism, game theory, and particle physics.

### Research-based fiction

We have already encountered the sociological fiction zine, *So Fi Zine,* which publishes short fiction on sociological themes. Its editor, Ashleigh Watson, studied the place of fiction in sociology for her PhD and became fiction editor of the Sociological Review in 2018. Other fiction has been written that was inspired by and based on research findings. For example, Patricia Leavy published a novel, *Low Fat Love* (Leavy, 2011) based on 10 years of interview research she had conducted with young women. Vicki Grant published another, *36 Questions That Changed My Mind About You*, based on psychological research from the 1990s (Grant, 2017).

If you want to write fiction based on your doctoral research, you will need some fiction writing skills. People often think writing is easy because they can read fluently but the two skills are very different. Fiction writing can be learned and there are many courses and writing groups available in most places and online.

## More alternative methods of publishing

This chapter has focused on the most common alternative methods of publishing from research, that is, zines, comics, web comics, graphic books, and research-based fiction. But there are other methods too. For example, some people have written songs from research findings. The song *Gwithian Sands*, by Kitrina Douglas, is based on the findings from 10 years of research about the lives and experiences of women in Cornwall aged over 60. It is an excellent example of the genre.

Another method is animation. Stacy Bias used this to great effect in her video *Flying While Fat*. This was based on her undergraduate research into the experiences of fat people travelling by plane. She surveyed 800 people and did in-depth interviews with 28, and her animation includes voice clips from those interviews (with participants' permission). The University of Liverpool helped to fund the animation.

Video itself is a form of publishing these days and many people have used that to disseminate research findings. One powerful example is *Have We Waited Too Long?* based on research on climate change conducted in Rigolet, Alaska, in 2010. There are other good videos from this project too, but the one we have selected is quite concise and full of information and meaning. The method used for the research was collaborative participatory digital storytelling.

Kip Jones from the University of Bournemouth went one step further and created a dramatic film, *Rufus Stone*. The film was based on the research he led into rural older people's isolation, connections, and sexual orientations. Funding was awarded by the Arts and Humanities Research Council.

All of the above examples are on YouTube and/or Vimeo at the time of writing. None are the only output from their research; they stand alongside, and enhance, research reports, academic journal articles, conference presentations, and other more conventional outputs. These innovative methods, and others, are available to you if you have the skills, the time, and the desire to use them.

## PIPs

**Kris** thinks alternative methods of writing and publishing are unlikely to be any help with their career plans at this stage. However, they write short stories in their spare time and their doctoral research generated an idea for a thriller with a tax inspector playing a central role. This is a persistent idea but Kris is dubious about whether they can find the stamina to create a long piece of prose fiction on top of all their academic writing. They have joined a weekly creative writing group held at their local library to try to figure out whether they really can and wants to write a novel. They are pleasantly surprised to find that what they learns from the group helps their academic writing too.

**Ella** loves the idea of zines and comics because she thinks they would be great ways to share her findings with teenagers, including her own participants. She would like to create a zine with and for her participants. However, she will need some funding. There isn't much funding available to independent researchers, but she has found one funding stream she is eligible for and is planning to send in an application once she has completed her doctoral study.

**Nathan** is aware of the growing body of health literature in graphic book form and is in contact with the *Graphic Medicine* network; he hopes to attend one of their conferences in the future. He has some skill in drawing and would like to create a graphic book to share his findings. However, he is aware of how long this will take, and isn't sure it's the most effective way to disseminate his work. He would relish the artistic challenge, but realises that it's not sensible to invest a great deal of his time in this when he has so much else to do. Therefore, he has decided to work on it in his spare time. He is writing the script first and, when that is finished, he intends to draw one page per week. He thinks that will be achievable and he should be able to get it finished within two years.

## Chapter summary

Alternative publication options allow you to com-municate ideas with visuals, audio, and/or other media. You have the potential to reach audiences that do not typically access scholarly publica-tions, or to complement formal publications with additional ways of presenting your ideas.

## Exercises and questions for reflection: planning and creating alternative publications

Use templates and worksheets found on the book website or www.path2publishing.com to complete this chapter's exercises and learning activities.

### Reflection

Which alternative formats do you enjoy most as a reader or listener? Why? How might your preferences contribute to your work with alternative methods of publishing?

### Exercises

*Part 1: research options*

1  Look at examples of the alternative methods of publishing research findings set out in this chapter that you find most appealing.
2  Identify content from your research that would suit at least two alternative methods of publishing.
3  Decide which alternative method of publishing is most closely aligned with your career goals.
4  Evaluate the resources (time, money, skills, etc) you would need to use that alternative method of publishing.

*Part 2: add to your publication strategy*

Write out goals and key steps. Include them in your publication strategy.

## Good practice points

- Make sure you have, or can acquire, the necessary skills to undertake an alternative publishing project.
- Don't take on a creative project unless you think you have enough capacity and resources to see it through.
- Audience: Think through your readership and which forms will reach them.

## References

Barker, M. J., & Scheele, J. (2016). *Queer: A graphic history*. London, England: Icon Books.

Duncan, R., Taylor, M., & Stoddard, D. (2016). *Creating comics as journalism, memoir & nonfiction*. New York, NY: Routledge.

Grant, V. (2017). *36 Questions that changed my mind about you*. London, England: Hot Key Books.

Leavy, P. (2011). *Low-fat love* (2nd ed.). Leiden, The Netherlands: Brill.

Sousanis, N. (2015). *Unflattening*. Boston, England: Harvard University Press.

# 9 Why, when, and how should I self-publish?

**After studying this chapter, you will be able to:**

- Understand motivations and options for self-publishing.
- Define key steps in the self-publishing process.
- Analyse ways to use your identified assets as the basis for self-published books.
- Evaluate implications for your publication strategy.

## Overview

Self-publishing covers a range of formats, for example, books, e-books, blogs, comics or graphic books, zines, podcasts, and audiobooks. Blogging is covered separately in Chapter 11, and comics, graphic books, and zines are covered in Chapter 8. This chapter will cover the other formats as well as more general issues around self-publishing.

Self-publishing is becoming mainstream in general terms, though at the time of writing it is still comparatively niche in academia. It requires a lot of hard work as well as offering scope for considerable reward, though this is by no means guaranteed. Some people find enough reward in having control of their own publishing decisions and schedule. And a very specific reward for some academics is being able to publish research findings fast, sometimes within days, rather than the months or years over which many commercial publishers operate.

There are many reasons for self-publishing. Perhaps you have a piece of 10,000–12,000 words. That's too long for a journal article or book chapter, but not long enough for a traditionally published book (even a short one, in most cases). However, e-books can be any length you like. Maybe you just want to write a short piece, 1,000 words or less, to make a particular point: That's perfect for a podcast. Or you might want to promote your ideas in an alternative format, to reach particular audiences such as car commuters or people with visual impairment, in which case an audiobook could be ideal. You could also decide to self-publish educational materials such as a case study, tutorial, workshop guide, or workbook. These materials could be used in conjunction with consulting activities or workshops, or complement your books or articles published commercially.

There is also a lot of flexibility about the content. From funny stories of your fieldwork to polemical rants, the choice is yours. Though if you really want to be taken seriously as a self-publishing academic, you will need to publish scholarly work and use some form of peer review. Maybe that will be an equivalent to formal peer review, with scholars in your field critiquing your work which you then revise before publication; maybe it will be more informal, such as by inviting comments on a podcast; or it could be something in between like finding beta readers for an e-book.

There are many decisions to make and a lot of roles to take on. You will be your own editorial, design, production, marketing, and sales operative. As this demonstrates, the work doesn't end with publication (although that also applies to com-mercially published work these days). You need to be prepared to market your work, because if you don't help your writing to stay visible, it can easily sink without trace. Don't expect that your brilliance will shine through all on its own: It won't.

So self-publishing is not for everyone. Maybe you're already thinking it's not for you. Do bear in mind, though, that there is scope to do just a small amount of self-publishing. One short e-book or podcast could attract attention from new audi-ences. We have both done this ourselves and found it was worthwhile. Therefore, even if you are not attracted to self-publishing at present, we would urge you to read this chapter. You never know when you might find the information useful.

## How to use your identified assets for self-publishing

Think about how the options listed in Figure 9.1 would fit with your publication strategy.

## When you might self-publish

You might self-publish because you want to:

- Write (or you have written) something that is hard to place with a commercial publisher because it crosses standard publishing boundaries
- Write (or you have written) something that is hard to place with an academic journal or book publisher
- Be independent of the commercial publishing business for ideological or other reasons
- Publish something quickly, such as time-limited research findings
- Reach a particular audience that prefers audio, media, or other formats
- Create a publication that is personalised and includes multimedia features not possible with typical print publications

# How can I use assets when self-publishing?

**Extract**

- Extract elements that fit the publication type.

**Condense**

- Summarize or edit sections that do not fit publication type.

**Expand**

- Add elements including visual or media elements.

**Adapt**

- Adapt your work to fit audience for the self-published book, podcast, or media.

**Apply**

- Make the self-published resource useful to readers such as a handbook, workbook, training guide or other application of research findings.

*Figure 9.1* How can I use assets when self-publishing?

- Develop cases, or ancillary or instructional materials that complement traditional publications
- Create how-to guides or handouts to give to your research participants or to people attending an event or workshop

Also, you might self-publish when you have time to create an output and can devote the necessary time to promote that output. Effective time management is one of the keys to successful self-publishing, not least because otherwise you could end up spending *all* your time creating, publishing, and promoting.

## The difference between self-publishing and vanity publishing

It shouldn't cost you anything to self-publish unless you are paying for the services of people such as text editors or brand/cover designers. Sadly, there are some so-called 'publishers' that will take writers' money in exchange for elaborate promises (Kara & Ryder, 2016). These are known as vanity publishers. Vanity publishing is not the same as open access publishing of journal articles with its author publishing charges (see Chapter 3 for details). Open access publishing moves the costs from readers to authors or their institutions. Vanity publishers charge authors to publish books that the publisher will then also charge readers to buy, so vanity publishers make money at both ends of the process. And some charge huge amounts. Four- or five-figure sums are not unusual for a 'publishing package' including several weeks of marketing, perhaps with promises of promotion in national newspapers, glossy magazines, or on prime-time television. These are likely to be empty promises and even if they do materialise, a few weeks of marketing is nothing for a book that you would hope will have a long shelf life. Books need sustained marketing over months and years.

Some vanity publishers present themselves as designed for 'self-publishing', but it's not self-publishing if you're paying someone else to do the whole thing for you. Genuine commercial publishers will not charge you for anything except perhaps the cost of the index, if you choose to have one made professionally (though non-fiction publishers should really bear the cost of a professionally made index, because an academic or reference book without a good quality index is not much used).

Vanity publishers are easy to spot because of their charges. Don't be deceived by appealing websites with glowing testimonials: If they want to charge you to publish a book, they're a vanity publisher. If you're still unsure, put their name into a search engine together with words such as 'reviews', 'complaint', and 'legal action'

(Kara & Ryder, 2016). This will show you whether there is any relevant information of the kind they would not be likely to put on their website. It only takes a few minutes and could save you a great deal of money and heartache.

## How to self-publish

It is essential to write something worth reading or hearing. Your creation should be accessible, informative, perhaps even entertaining. You may decide on the format first and then write to suit that format, or you may write something you're burning to write and figure out the format later. Exactly how you self-publish will depend on the format you choose and there is more information about this in the following sections. But whatever the content and format, your work must be good quality. Also, you will need to devote some time and effort – perhaps even money – to promoting your published work. If you're not prepared to do this, there is no point in self-publishing.

## E-books

Although self-publishing an e-book can be a solo endeavour, you might find that engaging others in the project will improve the quality and marketability of the product. To ensure that your e-book is readable, you will need help from at least two sets of people: peer reviewers and text editors. To ensure that your e-book is marketable you might also hire professionals who are experienced with cover design and formatting.

Peer review in self-publishing is not anonymous for the simple reason that you have to choose and approach your own reviewers. The best peer reviewers are people with good levels of relevant knowledge, who are able to give your work the time it needs, and can offer you thorough and constructive feedback. You may have to pay a professional text editor to help you when you reach that stage, unless you have a willing friend or family member with the necessary skills. Don't make the mistake of thinking you can edit your own work; none of us can do that. And don't make the other mistake of thinking that running your text through a spell checker is enough. Professional text editors look at everything from word choice to overall structure, and have the training and experience to identify problems and offer solutions. If you need to find a professional, ask around for recommendations, or go through an association such as the *Society for Editors and Proofreaders* (SfEP) in the United Kingdom, or the *Editorial Freelancers Association* in the United States.

(Note: Information and links for these and other services mentioned in this chapter are available on www.path2publishing.com.)

You can format an e-book yourself. However, there is a lot more to formatting than underlining your sub-headings. The platforms that distribute e-books have very particular requirements for the manuscript. If you want your e-book to be available on more than one platform, you will need to reformat the file to meet different sets of requirements. Alternatively, you could outsource this to an established and reputable organisation, with testimonials you can check, such as *Blot Publishing*.

The most commonly used e-book file formats are .mobi and .epub. Software is available to help you convert a Word file into the format used by the platform you select. However, there are other things to think about too. For example, you need to make sure that your e-book text is 'reflowable' (Kara & Ryder, 2016). This means that the font can be resized to make it accessible for readers with different qualities of eyesight; the text can be reoriented on screen from portrait to landscape presentation; and so on. Being reflowable makes e-books more user-friendly for readers. Unless you have good book formatting skills, the text will look better if it is professionally formatted, particularly if you include features such as images and diagrams.

An e-book needs more than just the contents: It also needs front matter, back matter, and a cover (Kara & Ryder, 2016). Front matter and back matter are the technical terms for the pages in between the front cover and the start of the book's content, and between the end of the book's content and the back cover. Readers often skip these pages but they are very important to publishers, though there are no hard-and-fast rules about what goes into them. Generally speaking, front matter includes a copyright statement, a title page, and a contents page. Front matter may also include a list of your other publications and/or a dedication. The back matter of e-books usually includes acknowledgements, and may also include a page asking readers to leave honest reviews online. Some writers include the beginning section of another book in their back matter in an effort to tempt readers to try that book too. If you want to see examples of front matter and back matter, look at a published e-book of the genre you want to create.

Covers are important marketing tools. They should be thumbnail-friendly so that they will look good on people's mobile phones as well as on larger screens. This means strong colours and not too much detail. There are three main ways to create covers: design them yourself, use a cover designer, or buy one off the shelf. To design a cover yourself you will need, at the very least, a good eye, some image editing software, and time to tinker. For some people this will be a straightforward task, for others it would involve a steep learning curve. If you're one of the latter, you will need to take an alternative route. With the increased popularity of e-books, more designers are focusing on this market. Ask for recommendations from published writers in the genre of your proposed book or look for examples you like and contact the designer.

You will also need to write a 'blurb', that is, a paragraph to help sell the book (Kara & Ryder, 2016). This is important even if you are making the book available for free, because you still need to give people a reason to upload and read your work. The blurb can be surprisingly difficult to write. It needs to be concise and yet explain what you have written and why it is worth reading. Emphasise the book's virtues, explain why you are qualified to write it, and say who makes up your intended readership.

E-book distributors help authors get published on multiple platforms, and distributed through diverse e-book sellers and libraries. The field of e-book distribution is evolving, and distributors come and go, but some of those that look most stable at the time of writing include *Draft2Digital, Smashwords, PublishDrive,* and *Streetlib*. There is also *KDP Select,* which is exclusively for Amazon and offers more ben-efits to offset the disadvantages of exclusiv-ity. These distributors take a small percentage of your profits in exchange for saving you the bother of re-formatting your e-book to meet the varying requirements of all the different platforms. If you are interested in this, you will need to research the platforms and their costs and benefits to decide whether it is a step you want to take.

ISBN stands for International Standard Book Number, a unique 13-digit number accompa-nied by a unique barcode that identifies a book across all platforms (ibid). Book titles can be replicated – there is no copyright on titles – so an ISBN is a good way for readers to make sure they are actually getting the book they want. However, they pre-date the digital era, and the counter-argument says that powerful digital search tools make it easy to find books using author names and keywords alongside titles. There is only one official source of ISBNs per country/region, and these agencies are all listed at   https://www.isbn-international.org/agencies. The agencies use ISBNs to track books and pro-duce industry statistics, which for some people is another good reason to use them. However, ISBNs cost money and the cost and speed of availability varies between agencies. In general, the more ISBNs you buy, the cheaper they are per unit – but they are not transferable, so you cannot sell them or even give them away. If you want only one, or a small number, they are likely to be quite expensive.

Some e-book publishers and distributors will give you an ISBN or another kind of identifying code. While identifying codes handed out by publishers or distributors are free, they only identify your book on that specific platform.

DRM stands for Digital Rights Management. This is an umbrella term for a group of technologies designed to prevent copyrighted works from being illegally changed and distributed (Kara & Ryder, 2016). They work by allowing content publishers to set restrictions on copying or viewing, which at first sight seems like a good idea. However, it is not quite so straightforward. Many big businesses use DRM technology and in most countries there are laws supporting the use of DRM. However, DRM is hard to enforce, can cause difficulties for legitimate customers (such as by locking them into a single bookstore), and may result in total loss of works if a specific DRM system – or its associated bookstore – becomes defunct. Also, there are many ways for people with enough technical know-how to get around DRM restrictions. Some people argue that DRM benefits big corporations, such as publishers, manufacturers, and software companies, far more than it benefits any individual author and copyright owner. So it's quite difficult to decide whether or not DRM is a good thing. Each author has to make their own decision on this, because you will be asked whether or not you want to enable it when you publish an e-book. We recommend that you research the latest thinking on DRM before you decide what works for your proposed project.

Pricing e-books is a topic of endless debate among self-published authors (Kara & Ryder, 2016). You can make your e-book available for free, price it low, price it average (if you can work out what that is), or price it high. A free e-book is easier to promote because people don't mind so much when you shout about a giveaway on social media. However, some people think free equals worthless.

To some extent price will depend on length, though there is not an absolute ratio. On the plus side, you can change the price of your e-book whenever you like, so if it's not selling well at one price you can give it at a different price. Some authors use 'price pulsing', which refers to reducing a price for a set period, for example, 24 hours or one week, and then raising it again. The suggestion is that promoting a 'special offer' helps to increase sales beyond the time in which the offer is available. Whichever methods you use to price and promote your e-book, don't expect it to make you rich.

In summary, questions to consider when self-publishing an e-book include:

* Which platforms will reach your target readership?
* Will you format and upload your e-book separately for each platform, or use an online distributor to deal with that side of things for you?
* Do you want to buy and use ISBNs?
* Do you want to enable DRM?
* How much will you charge for your book?

## Hard copy books

To self-publish a hard copy book, we would advise you to use a print-on-demand (POD) service. Some are linked with distributors while others stand alone. As with the distributors listed above, the field is evolving and POD publishers may come and go. The most stable at the time of writing appear to be *BookBaby, Blurb, IngramSpark, D2D Print,* and *KDP Print* (linked with Amazon). These companies are not all the same, they have different pros and cons, so do your research to make sure you're using the most appropriate option for you.

Even traditional publishers use a form of print-on-demand service these days rather than having to gamble on the size of a single print run before knowing how well a book will sell. This approach also works for self-publishing because you don't have to pay upfront for more copies than you need. Also, the quality is usually surprisingly good, and if anything does go wrong, POD companies are keen to correct it before you splash photos of their mistakes all over social media.

Covers for hard copy books are more difficult to create than those for e-books. This is because e-books only need a front cover, while hard copy books also need a back cover and a spine. You are more likely to need the services of a cover designer for a hard copy book.

## Podcasts and audiobooks

These kinds of publishing use sound rather than text. The distribution of podcasts and audiobooks has increased with the popularity of mobile devices that allow listeners to enjoy them anywhere and anytime. They are appreciated by people who can't read, or can't read easily, due to factors such as visual impairment or dyslexia. They are also useful for people who prefer to listen and take in information while they are doing other things such as commuting, exercising, or doing housework.

Podcasts are effectively DIY radio. They can be quite easy to produce using software that is available on most electronic devices, and can be placed online on a website or blog. A podcast can be any length you like, and can involve one person speaking, or two or more in dialogue or an interview format. The audience for podcasts is growing as more people find it's a format they enjoy. Podcasts can be distributed through channels and subscription services that serve as publishers for this type of content. In addition to news-oriented and topic-specific channels, there are channels dedicated to academic and scholarly podcasts that distribute to researchers and students globally.

Alternatively, you can create your own podcasts and distribute them from your own website or social media accounts. If you go the independent route you will be responsible for the technical steps involved, as well as marketing the podcast.

Audiobooks are more difficult to create, given that a book length typically exceeds that of a podcast. You need good quality sound recording and editing equipment, a quiet place to make the recording, and someone with a suitable voice. Many commercial audiobooks are voiced by professional actors; that works well but can be extremely expensive. Again, as an alternative to a totally DIY approach, you could work with an audiobook publisher and distribution service.

We are not advocating specific software, tutorials, etc., for making podcasts or audiobooks because technology and trends in this area change very quickly. If you want to publish in one of these ways, we suggest you do some research online to find out what people are currently recommending.

## Marketing

There is information about marketing in many chapters of this book but it is even more important when you self-publish. There is no marketing department to help you and without constant promotion your work will be invisible (Kara & Ryder, 2016). On the other hand, you can't spend all your time – and money – on marketing your work. You need to achieve a happy medium and only you can decide what that will be. You could measure marketing efforts in time, for example, two hours a week or half a day a month, or in activities, for example, two tweets per day and two Instagram posts per week.

Social media is useful if you can spend a few minutes once or twice a day and like interaction. Simply broadcasting 'buy my e-book' or 'listen to my podcast' doesn't work, you need to build relationships and only publish promotional posts now and again. If you can gather a community online you will have a ready audience for your work.

Mainstream media can also be useful if you have the necessary skills or are willing to acquire them. Face-to-face marketing is good too, for example, at conferences or workshops or when you're teaching. This doesn't have to involve any actual selling; you can just give out flyers or put your book cover image on your introductory PowerPoint slide. And you should always promote your work on your own webpage(s) or website(s).

## PIPs

**Kris** is not particularly interested in most self-publishing formats. However, they are a keen podcast listener. They have decided to make a pilot edition of a podcast for students in online courses and test it out on a few trusted friends. If their friends give it the go ahead, Kris will put the pilot online to do further market research with a view to developing a series.

They are looking at emerging podcast channels that focus on academic content, and hopes to find one that could distribute their work on a regular basis.

**Ella** is feeling torn about self-publishing. On one hand, she would prefer to work with an established publisher that would help with marketing and distributing her work. On the other hand, she is drawn to audiobooks, and creating them would suit her independent stance. She is considering the possibility of applying her findings in a practical audiobook that might reach families like those of her participants. Ella has found and bookmarked some webpages that give advice on how to make audiobooks and has written down her ideas. She's not planning to do anything further until she has her doctoral qualification, but feels pleased that she's created a resource for herself to use if she decides to take the next steps. She will include audiobooks in her long-range publication strategy.

**Nathan** wondered whether a free e-book might be a good way to make his findings more widely available, but he wasn't sure he could devote the necessary time to marketing. Then he had a brainwave and resuscitated his blog to record his post-doctoral journey. He realised he could use his community of 5,000 followers as an audience for podcasts adapted from his findings and their implications. He can't quite believe he didn't think of this sooner though he knows that the stress of finishing his PhD was the reason.

## Chapter summary

In the past, self-publishing was seen as a last resort for writers who were rejected by commercial publishers. Self-publishing was the domain of light fiction and personal memoirs. This is no longer the case, and all kinds of writers are taking advantage of independent and self-publishing opportunities for e-books or print-on-demand books. Writers with the skills and time can complete the entire project on their own. However, the popularity of self-publishing has in turn generated a range of services that writers can use for design or editorial tasks typically completed by commercial publishers.

## Exercises: planning and creating self-published works

Visit the Routledge book website and www. path2publishing.com for templates and worksheets you can use to complete this chapter's exercises.

### Reflective questions

What is your general view of:

- Self-publishing?
- People who self-publish?
- Self-published books?

How do you feel about self-publishing some of your own work?

### Exercises

*Part 1*

1 Think about your identified assets. Which elements would lend themselves to self-publishing, in which formats?
2 Which self-publishing format are you most attracted to? Why? Could publishing in this format really help your career goals, or would it be an end in itself?
3 Think about your target audience. What forms and formats do they prefer? Are there channels or subscription services that reach your target audience?

*Part 2*

Update the publication strategy you started in Chapter 2 to include the self-publishing options you are considering. Remember, you can find the publication strategy outline and other templates for download on the Routledge book site and www.path2publishing.com.

## Good practice points

- Create the best output you possibly can.
- Get help when you need it, for example, with editing, proofreading, or cover design.
- Other than this kind of help, don't pay to publish.
- Be prepared to market and promote your work.

## Reference

Kara, H., & Ryder, N. (2016). *Self-publishing for academics*. Uttoxeter, England: Know More Publishing.

# 10  Why, when, and how should I use social media?

**After studying this chapter, you will be able to:**

- Determine when and how social networking could be beneficial to your efforts as a writer.
- Create a strategy for smart use of social networking at each stage of the publication process.
- Select the most appropriate social media or online community sites.
- Think about how you can draw on your dissertation, thesis, and other academic work to create posts.

## Overview

This chapter offers a range of options for using online communities or commercial social media platforms to support your writing. Online social networking is increasingly important for academic and scholarly writers. We are both advocates for this, having received work and writing opportunities as a result of our online social networking that we never would have received otherwise.

However, we realise that some people don't like social media or social networking. There are all sorts of reasons for this. It may feel like one task too many, or people might have ethical objections to the way some social media sites operate, or it could feel futile to add yet another voice to the online clamour. We can't force you to network online, but we would strongly encourage you to give it a try. If you are disciplined about your time online, it need only take a few minutes each day, and that time can make a huge difference to the visibility of your published writing. But do remember that it is *social* media. Whatever platform(s) you use, it is very bad form simply to broadcast your wares without engaging with other people.

If you already have accounts on commercial platforms and are familiar with the ways they work, now that you want to be taken seriously as a writer, it is time to rethink how you use existing accounts and to decide whether you need new or different accounts. You might want to review and clean up student accounts, and it would certainly be worth investing some time to reflect on the identity you want to project online.

Once you've considered the options, integrate your social networking goals and plans into your publication strategy. Social networking can be a valuable approach to building your reputation as a writer, or it can be a complete waste of time (Zyga, 2017). Study this chapter and complete the exercises so you are prepared to make productive use of these communication channels.

## How to use your identified assets for social media

Social media posts are typically very short and incorporate images. These posts will often be designed to draw readers to other online or print publications. Use your identified assets to create posts with one or more of these approaches (Figure 10.1):

- Extract short bits, including definitions of terms or concepts, insights about the research problem, or findings. Also extract diagrams, visual maps, or other images suitable to post.
- Condense paragraphs to sentences or phrases.
- Expand social media posts with links to your online or print resources, or other relevant materials.
- Adapt work to fit social media site culture and format. Adapt complex scholarly work to simple statements or images suitable for general audiences.
- Apply ideas in usable steps.

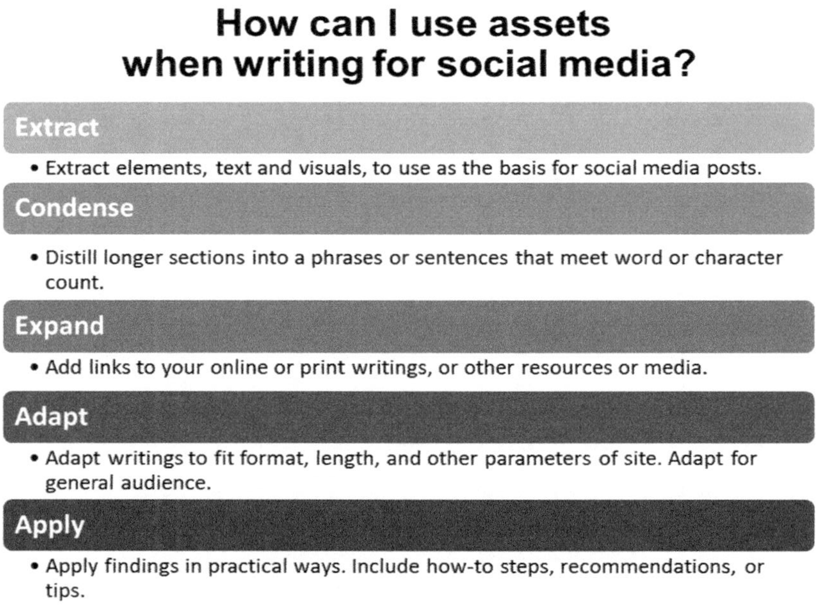

# How can I use assets when writing for social media?

**Extract**
- Extract elements, text and visuals, to use as the basis for social media posts.

**Condense**
- Distill longer sections into a phrases or sentences that meet word or character count.

**Expand**
- Add links to your online or print writings, or other resources or media.

**Adapt**
- Adapt writings to fit format, length, and other parameters of site. Adapt for general audience.

**Apply**
- Apply findings in practical ways. Include how-to steps, recommendations, or tips.

*Figure 10.1* How can I use assets when writing for social media?

## Defining terms

First, let's start by defining key terms. For the sake of clarity, we will differentiate between *social media* and *social networking*. We will also differentiate between commercial, public, or semi-private platforms, and private online communities. These distinctions are important to aspiring writers.

The term *social media* refers to websites, online platforms, or applications that allow for one-to-one, one-to-many, or many-to-many synchronous or asynchronous interactions between users who can create, archive, and retrieve user-generated content (Salmons, 2017). In social media, the user is producer; communication is interactive and networked with fluid roles between those who generate and receive content (Bechmann & Lomborg, 2013). Social media allows users to define and create groups, lists, or circles of 'friends' or 'followers' who have access to the user's content and can participate in text-based or visually enhanced text dialogue.

If the term *social media* refers to the online setting where we interact, *social networking* is the term we will use to describe the interactive activities we engage in on the sites.

We will use these terms throughout the book, but please keep in mind that as with many emerging technologies, the definitions have not been standardised. For example, others use *social media* to refer to the posts, podcasts, or other exchanges and *social network* to refer to the platforms (Mollett, Brumley, Gilson, & Williams, 2017). When collaborating with others, check your definitions to avoid confusion.

## Social networking and stages of your publication strategy

Think about social media usage in terms of your needs at all stages of the publication process: planning, writing, pre-publication, and post-publication.

### *Planning*

When you are planning a publication, you have specific questions. These could include:

- What should I write?
- Where should I publish?
- With whom should I collaborate?
- How should I proceed?

At the planning stage you want to learn from others. You can use social media to find others in your desired profession who have been successful, and observe them. How do they present themselves? What do they post, where, how often? You can find others with common interests in the questions you studied or the methods you used. You might find collaborators with whom you can write.

## Writing

Writing is challenging for most of us. It requires persistence and commitment to keep going. As a student you may have had friends with whom you could commiserate and faculty members who gave you feedback and advice. Writing on your own can be isolating, but social networking can help. Seek out other new writers; join an association, community, or group of writers; or participate in online events such as the annual Academic Writing Month (aka #AcWriMo) held each November. Once you've found or created a network, you can share ideas, questions, problems, or work-in-progress to crowd-source feedback.

## Pre-publication

Once you have a book contract or an article in review or accepted, your needs change. You still need the support of fellow writers, but now you also need to think about readers in a different way from when you were writing. How can you reach as many potential readers as possible? What substantive posts will attract their interest, so they will be ready to access or purchase the finished product? Can you cultivate relationships with influencers? Might they be willing to review your book for an appropriate journal, give testimonials, forward and share information about the book, article, or other publication?

### Publication

If you've found your social network during the writing and pre-publication stages, you will be ready when the work is published. At this stage, you can disseminate links to the article or promote sales of your book. You can use social media and online communities to create a credible identity as an expert in your field.

If you are working on several publication projects, you may be working at more than one stage at a time. You might be questioning where to go next at the planning stage for a new effort, while simultaneously promoting your latest publication. You could find that you need the writing support at all stages. You might also discover that one platform does not serve all purposes. As a student you may have just posted whatever was on your mind on a given day, or used social media to network with fellow students. You might have the kind of reputation fitting for a new academic PhD, or an online persona appropriate for student life. Now, careful thought about where to post what and when is essential. This is why you are looking purposefully at your options and making a plan as a key part of your publication strategy.

## Public versus private platforms for social networking

Once you have identified your needs at your current stage(s) of the publication process, you can think about what social networking activities will work for you on what social media site.

### Platform types

Commercial platforms include both public and semi-private types, while private communities have restricted access.

### Commercial public social media

Commercial public social media sites include the well-known brands such as Facebook, Twitter,

LinkedIn, and others. E-commerce storefronts or news sites where visitors comment or review products or articles also fit into this category. Typically, users post content with the expectation that it could be read by anyone. For some social media sites, such as Twitter, no registration or login is required to read posted material; to contribute a post, visitors will need to subscribe or register. For other sites, registration is required to read, post content, or comment to other users.

## Commercial semi-private social media

The opportunity to create closed groups is possible within many commercial social media platforms. Some of these groups are moderated or have "owners" who determine membership and participation norms.

While such groups give the impression that they are private, careful reading of the end-user licensing agreement (EULA) will show that posts made are kept private at the discretion of the platform owner. Any material posted resides on the company's servers. Situations have arisen where companies have been sold and interactions that users had perceived as private were made public. Also, members of seemingly closed groups are able to copy or share posted content.

## Private online communities

Educational institutions, companies, and professional societies set up platforms for social networking within their own groups. Sometimes called *walled gardens,* these spaces are not open to the public and require group membership as well as registration and passwords. These are members-only online spaces that require users to be an employee, student, or member to participate.

Platforms can include private forums, filesharing, wikis, and/or instant messaging features. Many of these platforms generate emails to subscribers when new posts come online, so members can readily stay involved in ongoing conversations. Users expect their comments to be private and available only to genuine community members. However, keep in mind that members can share posts or emails and avoid posting personal information you want to remain private.

## Selecting social media sites as a writer

Commercial platforms are designed for easy use and require little or no technical savvy. They operate for the unabashed goal of shareholder profit. Commercial platforms are not neutral spaces. They are free to users because they are supported by collection and sale of users' data to advertisers. The End User Licensing Agreements (EULA) spell out users' rights, including privacy and intellectual property rights. Users' rights can vary in different parts of the world, and these issues are far from being legally resolved.

The ease of use and popularity of commercial social media means writers can find others to network with professionally, as well as readers potentially interested in your books or articles. Writers who are concerned with the risks associated with posting material on commercial sites can decide to link to material posted elsewhere, rather than post their writing directly onto the platform. Use Table 10.1 to compare and contrast the options.

Private platforms can provide more protections for privacy and intellectual property, but they reach a restricted audience. Depending on the nature of your writing and the kinds of readers you hope to find, networking on the private platform set up for your professional society or group could be fruitful. In these spaces you can engage in more in-depth conversations and solicit advice from experienced people in your professional or academic field.

Ultimately, your selection of which social media sites to use and how to use them depends on the ways you answer the following questions:

- What stage(s) of the planning, writing, publishing, and promoting process will be the focus?

*Table 10.1* Researching social media options

| Platform | Pros | Cons |
|---|---|---|
| **Facebook** | • A Facebook page is useful for promoting your work<br>• Pages are quick to update | • Privacy and intellectual property concerns and ethical questions about the ways users' data are sold to third parties<br>• Algorithms mean you're unlikely to reach many people<br>• Not worth paying to increase reach |
| **Instagram or Pinterest** | • Use among academics is growing<br>• Great for visual thinkers<br>• 'A picture tells a thousand words'<br>• Short videos or still images can be posted<br>• Easy to use with a smartphone or mobile device | • Little room for text as needed to offer context |

*(Continued)*

*Table 10.1*  Researching social media options
(*Continued*)

| Platform | Pros | Cons |
|---|---|---|
| **Twitter** | • Great for international networking (except where blocked)<br>• Useful for promotion | • Can be a massive drain on your time<br>• Fast-paced: target audience may miss your posts<br>• Bots and detractors can create conflict |
| **LinkedIn** | • Good for promotion to special interest groups and professional networks<br>• Flexibility in length of post, can be used for blogging | • Less 'social' than most social media |
| **Private online community** | • Excellent for in-depth exchange with trusted colleagues or members of like mind | • Limited exposure to the general public |

- How do you want to use social networking, and with whom do you want to network? Are you trying to find other experts within your field or discipline, or are you looking for ways to reach the general public?
- Are you interested in sharing work-in-progress for feedback?
- Are you trying to promote completed work to encourage readers and/or buyers?
- Is your work of a sensitive or proprietary nature?

## Posting your writing on social media sites

Once you have selected the site(s) best aligned with your goals, look at the audit you did in the Chapter 1 exercises and reflect on ways to use your identified assets as the basis for short material suitable for posting. To get started, consider what you hope to accomplish with social networking. Here are some options:

- If you are trying to find others of like mind, you could extract key statements that illustrate your values and interests.
- If you are trying to introduce yourself and your work, you might want to condense your research design, purpose, and results into a concise statement you can post.

- If you want to reach an audience beyond academia, you could expand on your presentation by creating a short media piece to post.
- If you want to attract attention, you could adapt part of your writing into a visual infographic or create a short media piece to post.
- If you want to reach an audience beyond academia, you could create materials to help them apply what you've learned.

## PIPs

**Kris** is at the planning stage. Their first step is to look at existing social media accounts, rewrite profiles to reflect post-student status, and check for any potentially embarrassing posts that do not align with the academic reputation they are working to establish. Kris will also examine the potential in each platform for the kind of support and exchange they anticipate throughout the writing process.

Kris realises that the commercial social media sites they enjoyed in student days can be useful in some ways, but they need a community now that they are trying to establish scholarly credibility. Kris decides to join a professional association in the field where they want to land an academic position, and get involved with a special interest group (or SIG) that has an active members-only discussion forum online. Kris will look at the audit completed in the Chapter 1 exercises. They will select statements, quotes, or observations from their identified assets to extract or condense into meaningful posts.

Like Kris, **Ella** will start by updating existing social media accounts. Her first social networking goal includes finding places where LGBTQIA+ parents discuss issues, problems, and concerns. She will interact with them and try to understand their needs, so she can determine the focus for her future book. Where appropriate, she will adapt resources or findings from her doctoral writings into posts that will be of interest to parents.

Ella's second social networking goal involves learning about the journal submission process. She plans to follow published writers, editors, and journals on social media to glean what she can about success strategies. She is already planning to participate in the first Academic Writing Month after her doctoral study is completed, and intends to have her article outlined by then. She hopes that she will find new friends who are committed to academic writing whether they are working inside or outside of the academy.

**Nathan** has a clear concept for his first publication, so he feels he is at the writing and pre-publication stages. He does not want to dedicate a lot of time to social networking on commercial social media sites, because he already has a professional network of people he trusts plus his blog followers. He plans to use social media sites to look for opportunities to promote his publications, and only to do as much networking as necessary to keep his accounts from looking like they're broadcast-only.

## Chapter summary

Social networking can be beneficial or detrimental to writers. Commercial social media is notorious as a time-waster and distraction from the tasks at hand. At the same time, social networking can allow us to find others of like mind, or others with fresh perspectives, who help to enlarge our thinking. By thinking through how and where we want to participate in social media, we can purposefully align our posts and activities with stages of the publication process.

## Exercises and questions for reflection: planning and networking with social media

Use templates and worksheets found on the Routledge book website or www.path2publishing.com to complete this chapter's exercises and learning activities.

## Reflection

How does your social media presence need to change as you move beyond student life? How does your personal usage of social media fit (or not fit) with your plans for professional life and authorship? Think through any concerns.

## Exercises

### Part 1: social networking and online communities

Social media websites and online communities offer many opportunities for publication. They are also very useful for dissemination and promotion. In some form or another, social media should be part of your publication strategy. If you don't use social media at present, evaluate options and choose the sites that are popular with your target readers. This exercise will help you to take a strategic approach to your online social strategy.

A. Locate and review online communities and social media pages created by institutions, organisations or associations, or individuals in your field or discipline.
B. Use Table 10.2 to complete an audit of your social media presence. Label the platforms you want to evaluate.

Table 10.2 Planning social media strategy

| Platform | Profile completed, including photo, and all up to date? | Number of followers or equivalent | Level of your activity (daily, twice daily, weekly, monthly, less frequent, etc.) | Specific changes you would like to make (e.g., complete profile, increase followers by 10% in six months, post topical content each morning) |
|---|---|---|---|---|
| Facebook (page/group, not personal account) | | | | |
| Twitter | | | | |
| Instagram or Pinterest | | | | |
| LinkedIn | | | | |
| Members-only professional community | | | | |
| Other | | | | |

C. Take this three-step approach to revising and refining your 'brand':

1 If there is another platform or platforms you would like to use beyond those you already use, set up a profile or profiles there.

2 Assess and align your profiles across all the social media you use. This does not mean making all your profiles identical (which wouldn't be possible anyway due to space and other constraints). It means making sure each profile tells a similar story in a similar way to the others.

3 Make sure you are using the same photo on all the social media you use, plus other online functions such as Skype and Google.

Use the information you have identified about the changes you want to make to help you develop your social media strategy for the next 6–12 months. Do this by completing Table 10.3. Be realistic about the time commitments you intend to make. You don't need to spend a long time using social media: an investment of as little as 10 minutes per day, used strategically, can pay considerable dividends (and that can be, say, in $2 \times 5$-minute chunks morning and evening).

*Part 2: publication strategy*

Use your work from the Part 1 exercises above to add goals and plans to your publication strategy.

*Table 10.3* Using social media as a researcher

| Platform | When will I post? | What will I post? | When will I respond to others' posts? | What will I do to increase engagement with my work? | How will this benefit my career? |
|---|---|---|---|---|---|
| | | | | | |
| | | | | | |
| | | | | | |

## Good practice points

- Think about your academic and professional identities and the ways you want to present yourself on social media.
- Where possible, delete past posts that are counter to your academic and professional social media goals.
- Determine the best approach for introducing yourself in your post-student persona.
- Be consistent with posts that complement and amplify other publications.

# References

Bechmann, A., & Lomborg, S. (2013). Mapping actor roles in social media: Different perspectives on value creation in theories of user participation. *New Media & Society*, *15*(5), 765–781. doi: 0.1177/1461444812462853.

Mollett, A., Brumley, C., Gilson, C., & Williams, S. (2017). *Communicating your research with social media: A practical guide to using blogs, podcasts, data visualizations and video*. London, England: SAGE Publications.

Salmons, J. (2017). *Teaching ethics & social responsibility in a conflicted world*. Paper presented at the Academy of Management Annual Meeting, Social Issues in Management Division, Atlanta.

Zyga, K. (2017). Focus groups: Insight consulting group.

# 11  Why, when, and how should I contribute to a blog?

**After studying this chapter, you will be able to:**

- Identify types of academic blogs and their purpose.
- Consider ways to include blogging in your publication strategy.

## Overview

Academic writing is often a long and complicated process. But maybe you just want to write a short piece, 1,000 words or less, to make a particular point: That's perfect for a blog post (or, as we saw in Chapter 9, a podcast). In this chapter you will explore types and uses of academic or research-based blogs, and reflect on ways they could complement other approaches included in your publication strategy.

Blogs were popular in the early days of the web, then fell by the wayside when social media platforms were in favour. (See below for an explanation of the difference between blogs and social media.) Over time, writers have found that many social media, particularly commercial platforms, had significant limitations in terms of format, style, and length of posts. Academic writers in particular want more control over how their work is presented. As a result, academic blogging is having a resurgence.

Today's researchers, scholars, and instructors are finding many valuable ways to use blogs. Academic writers are using blogs in conjunction with formal publications such as books and articles, and informal posts on social media. In addition to communicating with other academics or professionals, researchers can use blogs to provide information about their studies in ways that would be useful and interesting to prospective and current participants. Blogs can serve as a bridge between researchers and participants.

## How to use your identified assets for blogs

Blog posts can vary in length and format. While blogs are usually available to anyone online, they can be focused within specific scholarly or professional communities. Use your identified assets to create posts with one or more of the approaches suggested in Figure 11.1.

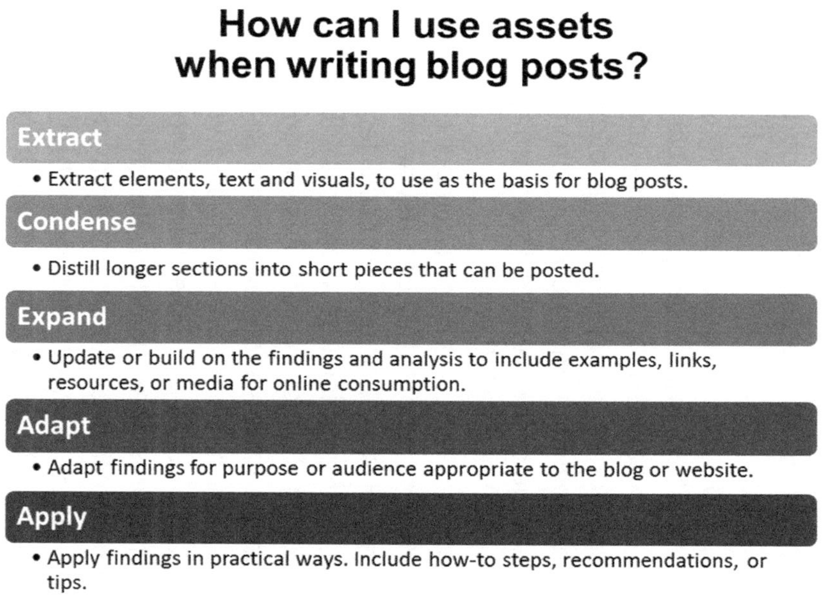

# How can I use assets when writing blog posts?

**Extract**
- Extract elements, text and visuals, to use as the basis for blog posts.

**Condense**
- Distill longer sections into short pieces that can be posted.

**Expand**
- Update or build on the findings and analysis to include examples, links, resources, or media for online consumption.

**Adapt**
- Adapt findings for purpose or audience appropriate to the blog or website.

**Apply**
- Apply findings in practical ways. Include how-to steps, recommendations, or tips.

*Figure 11.1* How can I use assets when writing blog posts?

## Blogs and social media

Before going any further, it's important to define blogs and to distinguish between blogs and social media (see Chapter 10 for more about social media).

Blogs, short for weblogs, are a form of online publishing where entries are posted chronologically and organised by categories and tags. Broad categories and specific tags allow readers to find posts that may no longer be visible on the front page of the website. Posts are typically between 500 and 1,000 words.

Bloggers can choose from a number of free or paid platforms where they can devise their own templates, or adapt or use available templates. Popular platforms include WordPress, Blogger, and Medium. Blogs are very flexible and users can create a wide variety of formats and styles of presentation. Some are very basic with simple narrative posts and others are complex with design features that include both static pages and time-sensitive posts. Bloggers can use comment features to invite feedback or to interact with readers. Bloggers may choose to generate revenue with advertising and other promotions.

Bloggers are not limited by prescribed lengths, styles, features, or page designs. This flexibility stands in contrast to posts made to social media sites. Social media sites are typically run as commercial platforms by large companies. These companies have determined ways to make a profit from user-generated material and are thus invested in allowing certain kinds of posts. Brands are built on the sites' graphic design and features.

Users of sites like Facebook can be surprised to log in and find that their pages' format has changed without their knowledge or permission. They have become accustomed to seeing advertising on their walls, as well as links to other content the company has decided is of interest to users who fit a certain profile. Users of the microblog tool Twitter have become accustomed to the 280-character limitation current at the time of writing (formerly 140 characters; this may change again).

Social media sites and blogs are typically interconnected (Table 11.1). Bloggers use social media to build an audience. They create posts that fit within the constraints of the social media sites, but link back to the blog where they have the freedom to present information in the way they prefer. As we saw in the previous chapter, social media refers to the online setting where we interact, while social networking is the term we use to describe interactive activities on the social media sites. Bloggers can use social networking to interact with others and use their blogs to present more substantial writings and

*Table 11.1*  Blogs and social media sites (Salmons, 2018)

| Blogs | Social Media Sites |
| --- | --- |
| Flexible formats and options for presenting narrative material of any length, attachments for download, graphics, photographs, and/or media. | Format options determined by commercial owner of site. |
| Communication features determined by blogger, using free, open-access plug-ins and software or professionally designed templates. | Features and design options determined by commercial owner of site. |
| Advertising determined by the blogger, who collects any revenue. | Advertising determined by the commercial site, and the site collects the revenue. |
| The blogger chooses what content to promote and what links to share. | Commercial owner of site uses data analytics to select content and links aligned with visitors' interests. |

other expressions. The important point here is that by knowing the strengths and weaknesses of each option, you can make better choices about which best meets the goals you have articulated in your publication strategy.

## Scholarly blogs

Scholars can use blogs in three main ways (Salmons, 2018):

- **Researcher-to-Researcher:** sharing and exchange
- **Researcher-to-Participants:** building credibility and 'informing' participants
- **Researcher-to-Public:** sharing findings, results, and practical resources

**Researcher-to-Researcher:** Bloggers in professional contexts communicate with each other for exchange and networking within or across disciplines or fields. They present information in ways that build upon a shared foundation in the topics under discussion, and a shared understanding of the protocols and expectations for activities such as conducting research or teaching at the college level. They share resources, links to recent publications, calls for papers, or notices about upcoming conferences, or other opportunities of interest to other academics. Researchers use academic blogs to try out ideas and solicit feedback, before refining their writing for more formal publications (Carrigan, 2020).

**Researcher-to-Participants:** Researchers communicate with prospective or current participants. They can present information in ways the study population will understand. Blogs intended for this purpose can introduce the study, develop the credibility of the researcher, be used to support recruitment efforts, and help to inform participants before and during the study.

**Researcher-to-Public**: Research bloggers also communicate with the general public. They present information in ways that are interesting and understandable for anyone. The researcher may translate academese or disciplinary jargon into more familiar terms. They may offer recommendations for applying findings in practical ways. This type of blog is designed to share findings and/or to build awareness about the issues and problems under investigation.

Patrick Dunleavy (2014), focusing on academic blogs, introduced another way to categorise blogs. He distinguished between them according to the number and type of author: solo, collaborative, or multi-author (Dunleavy, 2014). Table 11.2 is adapted from his model.

The columns are intended here not as a fixed set of boundaries but more as a continuum between, on one end, the individual DIY blogger who is responsible for everything from choosing the domain name, platform, hosting service, and figuring out how to make it all work to, on the other, a professional operation more comparable to a journal or magazine production.

Table 11.2 Scholarly blogs and audiences

| | Solo | Collaborative | Multi-Author or Institutional |
|---|---|---|---|
| **Researcher-to-Researcher** | The researcher creates a blog where he or she posts information about research interests, projects, conference presentations, and publications. | Researchers with a shared interest, area of inquiry, methodology, or discipline work together to create a blog about their individual or team research projects and related events and resources. The blog may serve as a channel for connecting with new research partners, conferences, or funding opportunities. | Professional society, association, or university group sponsors a blog for researchers working in a specific area of inquiry or discipline. Writing in the blog and any related publications is aimed at other scholars and professionals in the field. |

*(Continued)*

*Table 11.2* Scholarly blogs and audiences (*Continued*)

| | Solo | Collaborative | Multi-Author or Institutional |
|---|---|---|---|
| **Researcher-to-Participants** | | Individual researchers or research teams use a blog to explain the purpose of the study and expectations, benefits and/or risks for participants. As appropriate, findings are shared with the participants. Links are shared on social media sites. | Announcements for opportunities to participate in studies are posted on the blog. Links are shared on social media sites. |
| **Researcher-to-Public** | The researcher creates a blog where he or she posts information about research findings and | The group of collaborative researchers and writers creates a blog with the intention of disseminating research findings to those who can use them. The researchers may use the | Articles about application and practical use of research findings are featured on the organisation's blog. A blog may be one |

their application. The researcher may use the blog to promote his or her workshops or consultations about how to apply research findings. Links are shared on social media sites.

blog to promote workshops or consultations about how to apply research findings. Links are shared on social media sites.

of many channels for reaching the public. Articles may be associated with products and services available for sale such as handbooks, workshops, or training. Links are shared on social media sites.

## How can I start blogging?

You can design and maintain your own solo blog, find a group of like-minded writers who want to begin a collaborative blog, or look for opportunities to contribute to a multi-author blog in your field or related area of interest. There are pros and cons to each of these approaches. Starting your own blog enables you to showcase your work. The disadvantages to this are that blogs require regular content and it takes time to build an audience. You may be brimming with ideas right now, but can you put in the sustained effort needed to write at least one carefully crafted 800-word post per week for several years? And promote your own work? That is what you will need to do if you want to gather your own audience. In collaborative blogging the writing and promotional work is shared, which can mean less work overall. However, you would need initial discussions to establish who will do what and how to manage coverage for holidays, sickness, etc. To contribute to a multi-author blog, you need to craft your post even more carefully than you would for your own blog, and the editor may still send it back with a request for corrections. On the plus side, these kinds of blogs already have an audience, and they are likely to do at least some of the promotion for you too.

You might find that new partnerships, skills, technologies, and/or services are needed to move forward as a blogger. For example, you will need to learn how to find and use a blogging platform if you plan to operate a solo blog. If you decide you want to include media, or audio podcasts, you may need to purchase a camera or learn how to capture clean recordings. These implications might influence your choices about whether you want to use free or paid services, or whether you decide to contract with others to take care of some of the administrative or technical tasks.

## Writing a blog post

The ideal length for a blog post is 700–800 words, though they can be shorter or longer. A post should cover a single issue or make a single main point, though obviously this can be done in some detail. Nevertheless, there is no excuse for waffling. You should write concisely within a clear narrative arc. A good blog post is often structured like a short story, with a beginning, a middle, and an end.

Each blog post also needs at least one image or other visual element, sometimes several, perhaps including embedded video clips. Be sure to use your own images or source them from a reputable royalty-free site such as Pixabay to avoid copyright problems. Also, use hyperlinks where they might be useful for your readers. They can be useful for you as a writer, too. For example, if you mention an organisation, you can link to that organisation's website rather than describing it. Then readers who don't know of the organisation, or who want to know more about it, can click through to find out.

## PIPs

**Kris** might be interested in contributing to a multi-author or institutional blog associated with their field. They can see how informal interaction with others in the field could be stimulated by blog posts about the research. However, as they are focused on a career in academia, journal articles are most important for them and starting their own solo blog is not a priority.

**Ella** is enthusiastic about the idea of creating a collaborative researcher-to-public blog, about LGBTQIA+ families, with contributions by scholars, parents, teachers, social workers, and others. This kind of blog will be important to her long-term publication and professional strategy as an independent scholar. But since she has more immediate priorities until her graduation, she will start by looking for opportunities to create guest posts to reach her target audience. This will give her a chance to develop her skills as a blogger. She intends to extract and adapt sections of her thesis for blog posts that will introduce her work.

**Nathan** wants to build a career as a psychiatric nurse researcher for a research institute or government agency. As well as reviving his own blog, he plans to explore the institutional blogs operated by the kinds of places he wants to work. He will examine the types of posts made to determine whether they are aimed at researcher-to-researcher exchange, researcher-to-participant information, and/or researcher-to-public dissemination. Once he has a better understanding of blogging in his field, he will decide where he can best contribute and update his publication strategy accordingly. He hopes to find a blog that meets one or more of his target audiences (mental health workers and military families) where he can write guest posts. He intends to condense writings about the premise and purpose of his book, and create posts for relevant blogs that he hopes will generate interest.

## Chapter summary

Academic blogging is fast becoming a recognised way to disseminate findings, build networks, share effective practices, and try out new ideas. You can use blogging, in conjunction with formal publications, and in coordination with social media, to create a publication eco-system that helps you to achieve your goals. While you can use free or inexpensive plat-forms to create your own blog, you can also join with others or offer guest posts on blogs run by academic institutions, professional associa-tions, publishers, or related organisations.

## Exercises and questions for reflection: planning and writing for blogs

Use templates and worksheets found on the Routledge book website or www.path2publishing to complete this chapter's exercises and learning activities.

### Reflection

What purpose do you hope to achieve by start-ing a blog or contributing to a blog?

### Exercises

*Part 1: planning exercises*

**Exercise 1:** Find at least three different academic or research-oriented blogs by and/or for people in your field, discipline, and/or profession.

- Compare and contrast the content and styles of these blogs. Which do you like best and why? What do you dislike and why?
- Describe the features these blogs use to attract and retain readers.

**Exercise 2:** Using your dissertation or thesis as the basis for blog posts: Think about your thesis or dissertation and other academic writings, and review the audit you conducted for Chapter 1. Is there a part of your work that would lend itself well to blogging? Create a table like Tables 11.3 (Example 1) and 11.4 (Example 2) to organise your ideas.

Research options for making a guest post. Which bloggers might you approach? Why? How and where would you publicise a post?

**Exercise 3:** Research options for creating your own blog.

- Which platform would you choose? WordPress? Medium? Blogger? Other? Why? Free or fee-based? What is the rationale for your choices?
- How often would you post?
- Would you include guest posts from other writers?
- Would you include links to other online content? If so, describe.

*Table 11.3* Blog strategy example #1

| | Part of Thesis or Dissertation | Extract, Condense, Expand, Adapt, or Apply? | Alignment With Goals |
|---|---|---|---|
| **Guest post** | Use introduction to the problem from Chapter 1 as the basis for a blog post. | Extract | Build credibility for my understanding of the issues. |
| **Own blog** | Use recommendations for practice as the central theme for a blog aimed at practitioners. | Adapt and expand | Improve research impact |

- Would you consider advertising or product placement on your blog?
- How and where would you publicise your blog?

**Exercise 4:** Write an 800-word blog post. Include images and hyperlinks.

**Exercise 5:** Update the publication strategy you started in Chapter 2 to include any blogging activities you think will fit.

Table 11.4 Blog strategy example #2

| Option | Description | Related Goals | To-Do List |
|---|---|---|---|
| 1. Extract | Use introduction to the problem from Chapter 1 as the basis for a blog post. | Build credibility for my understanding of the issues. | See whether my professional association has a blog where my post might gain attention from others in my field. I might ask whether anyone would like to collaborate on an article. |
| 2. Condense | | | Look for a journal with a global audience. Also, look for calls for chapters in books from top-tier publishers that might have broader distribution. |
| 3. Expand | | | |
| 4. Adapt | | | |
| 5. Apply | | | |

## Good practice points

- Create blog posts that are succinct but substantive.
- Complement, but don't duplicate, other published writings.
- Link to related resources or media.
- Use social media to draw attention to blog posts.

## References

Carrigan, M. (2020). *Social media for academics* (2nd ed.). London, England: SAGE Publications.

Dunleavy, P. (2014). Shorter, better, faster, free: Blogging changes the nature of academic research, not just how it is communicated.

Salmons, J. (2018). Research impact and academic blogging. *Research Information*, 22–23.

# 12 How can I implement my publication strategy?

**After studying this chapter, you will be able to:**

- Refine the publication strategy you have developed through the series of exercises recommended in Chapters 1–11.
- Create a timeline for your publication strategy.
- Develop a system of accountability to help you stay on track.

## Overview

In Chapter 12 we invite you to tie together the pieces you have explored throughout the book. Now is the time to finalise a thorough and flexible publication strategy that fits with your life and career goals and makes the best use of the research and writing you have completed during your doctoral studies.

A strategy is useless unless it is paired with an action plan. This plan is likely to need regular review, at least once a month, due to circum-stances beyond your control. For example, academic journal timescales are quite unpre-dictable, partly because peer reviewers take varying lengths of time to respond. In the mid-dle of a well-planned month you might have an email from a journal administrator with reviewers' comments, requesting amendments to be made within a comparatively short timescale. Or, like both Helen and Janet, you might take a firm vow, one morning, never to write another book chapter ever again, and then get a call from a friend at lunchtime that day, asking you to write a chapter for a collection he's editing – and you don't have the heart to refuse. So, you want a plan that is flexible enough to bend to fit new demands or lucky breaks, yet firm enough to keep you on track.

Writing is often a solitary activity. Typically, no one is looking over your shoulder to see whether your computer screen shows social media sites or a document in progress. We encourage you to find other writers who can offer mutual sup-port and encouragement.

Depending on where you studied, as a stu-dent you might have operated within an aca-demic term, under strict deadlines set by faculty or targets agreed to with supervisors. After you graduate, it is up to you to create timelines and stick with them. We can promise ourselves we will follow through to enact the plan but that might not be enough. Life can get in the way, particularly if you have started a new position after graduation. Use this chapter's exercises to develop an accountability system that will help you stay on track.

## Publications and purpose revisited

Think about who you want to reach with your thoughts, ideas, and research findings. Think about the impact you want your research to have – including influencing other scholars and your field, or providing new understandings or practical help to those who work in the areas you have studied. Now, revisit the steps you've taken in the recent weeks and flesh out your publication strategy.

In the reflections and exercises from Chapters 1 through 11 you considered characteristics of each type of publishing and looked at career-related benefits. You carefully analysed all elements of your dissertation or thesis, as well as other papers, reports, research notes, and memos. You identified your assets, the term we have used to describe particularly strong or insightful pieces of writing, and you decided what kinds of publications to aim for. However, you would need superhuman powers and more than the 24-hour day to do them all! That is why you need a strategy that allows you to select the appropriate options, and stay focused on the publication options that will be most significant given your goals.

## From purpose to plan to action

Creating a realistic, achievable publication strategy means prioritisation. In order to set realistic priorities, you will need to reflect on your purpose for publishing. A sense of purpose is a strong motivating factor for each of us. A belief in the value of what we can contribute may keep us going through times of self-doubt, rejections, or the simple distractions of everyday life.

Let's return to the three uses of research findings identified by Rallis and Rossman (2012) and align them with ways to use your identified assets and with your publication purpose.

The first option Rallis and Rossman (2012) identified is *instrumental use*, which means research findings or other insights are used to solve a problem. This use of research depends on your ability to apply what you learned from the research experience or discovered from research findings. You might need to adapt scholarly language to more practical and succinct descriptions. You might also need to expand on some points to make them relevant in other settings.

Instrumental use can mean research impact, bridging the worlds of academia with the world of individuals, families, groups, organisations, or societies with real problems. Instrumental use can fulfil purposes such as changes to policy or practice. If your study explored theory, methodology, or methods, instrumental use could include developing new approaches other scholars can adopt to improve their own research.

Writers with an interest in instrumental use might look beyond academic journals and publish their work in books, professional journals, or on blogs or social media that reach the kinds of readers who can use the ideas to make a difference.

The second option Rallis and Rossman (2012) identified is *enlightenment use*, which means findings contribute to a deeper understanding or insight that over time can inform decisions or actions. Enlightenment use suggests a scholarly contribution to new knowledge about the topic or problem that was investigated. This use of research depends on your ability to extract significant sections from your identified assets or to condense a large document into an article-length piece.

Writers with an interest in enlightenment use of their research might be focused on scholarly journals or books. Such writers might contribute excerpts or commentaries to academic or discipline-focused websites or blogs hosted by universities, university presses, or professional societies.

The third option Rallis and Rossman (2012) identified is *symbolic and political use*. This option goes a step beyond instrumental use, to focus on the kinds of research impact associated with transformation at the individual level or action, justice, or change at the societal level. Political uses of research shape legislative or policy developments. To use research in this way, you can extract key background information about the problem or selected findings, and adapt them for the purpose at hand. You can also apply findings by articulating practical steps needed to put research results to use.

Writers in this category may find that the publication options likely to help them achieve their purpose will be outside traditional academic channels. Writers who want findings on fast-moving topics to have symbolic or political impact might need to work with publications that reach their readers in a timely manner. They might also need to reach a wider audience of people who feel the effects of the problem, or people advocating for change. Blogs, self-published e-books or how-to guides, as well as alternative types of publications, should be considered.

## Using and developing resources

To achieve the desired purpose with your publications you may find that you need extra resources.
These could come from new partnerships, skills,
technologies, and/or services. For example, if
you decide to begin blogging, you may need to
learn how to find and use a blogging platform. If
you decide you want to include media, or audio
podcasts, you may need to purchase a camera
or learn how to capture clean recordings. Do you
have a budget with funds you can allocate to a
new project? This might influence choices about
whether you want to use free or paid services,
or whether you decide to contract with others to
take care of some of the administrative or technical tasks. If money is tight, can you set up a skills
or equipment swap with someone?

## Can it be done? Strategies and success stories

You have seen the progression of the PIPs in
Chapters 1–11, which were designed to illustrate
how each type of publication might fit, or not,
depending on an individual's field and career
goals. To complete this stage of their journeys,
tables 12.1, 12.2 and 12.3 show the decisions
about publication strategy that they've each made:

Table 12.1 Kris's strategy

Highlights of Kris's publication strategy include

| Publication Steps | Timeframe for Completion | Accountability |
| --- | --- | --- |
| Look at social media accounts, delete student pictures, create profile appropriate for review by search committee. | Two weeks | • Set up a project plan using an electronic calendar, set up alerts for checkpoints<br>• Join **a professional association,** and participate in a Special Interest Group (SIG) for early career researchers<br>• Participate in a paper development workshop; look for others who might want to collaborate or just share ideas |
| Explore journals in both the subject matter of their study and research methods. Look for notices about relevant special issues. | One month | |
| Post on social media sites hosted by professional association where they are a member. Build network. | Ongoing | |

| | |
|---|---|
| Identify two to three journals for deeper dive. Review submission guidelines. Read articles that are similar to articles they would like to write, review closely to identify style, organisation, structure. | Two months |
| Select the first journal to approach. Create a plan for completing the article to meet submission deadline. | One month |
| Plan and write an article. | Six months |

Table 12.2 Ella's strategy

*Highlights of Ella's publication strategy include*

| Publication Steps | Timeframe for Completion | Accountability |
| --- | --- | --- |
| Work on journal article with doctoral supervisor. | Three months | • Create a chart with key steps and due dates to post on her home office wall |
| Clean up social media accounts to remove student-oriented posts. Find and join members-only groups where LGBTQIA+ issues are discussed. | Ongoing | • Contact former student friends, people in her professional network, to find two to three people interested in LGBTQIA+ youth who want to form a writing group for mutual feedback and encouragement |
| Research publishers of cases related to LGBTQIA+ youth. | Two months | |
| Follow blogs, podcasts, social media about academic and professional writing. | Ongoing | |
| Plan and write a case study. | Three months | |
| Research and develop a proposal for funding a zine. | Two months | |

Table 12.3 Nathan's strategy

Highlights of Nathan's publication strategy include

| Publication Steps | Timeframe for Completion | Accountability |
|---|---|---|
| Research publication options for a practical guidebook; initiate contact with acquisitions editor. | One month | 1 Create a detailed timeline. Ask partner/close friend to help him stay on track. |
| Develop a book proposal. | Two months | 2 Contact former student friends, people in his professional network, to find two to three people in the mental health field who want to form a writing group for mutual feedback and encouragement. |
| Write an article for open-access journal. | Six months | |
| Start writing regular blog posts. | Ongoing | |
| Record short podcasts to complement blog posts. | Ongoing | |
| Create cases to include in the book. | Three months | |

## Beyond fiction: PhDs who are implementing their strategies

Let's look at examples from real-life PhD graduates who have used the typology discussed in this book, and published articles, books, and other materials. The chapter numbers they refer to equate to the chapters of the typical thesis or dissertation as set out in Chapter 1 of this book. *These students started to think about future publications while still conducting their doctoral research. Do any of their experiences point you towards new thinking about your own next steps?*

### *Lisa Toler, PhD*

My post-PhD career goals were two-fold. I first wanted to expand my professional career in management. I was successful in this as I was promoted within a year after achieving my PhD in order to lead a team of professional, administrative, and scientific staff. Second, I wanted to teach and was hired as an Associate Online Adjunct Professor within eight months of receiving my PhD. I am not sure if doors were opened initially for me due to my publications. Those doors were opened mainly due to achieving the Phd. I do, however, believe publications have strengthened my professional value as an expert in the field as well as growth potential both as a project management professional and manager working in a scientific community. Publications have also strengthened my opportunities to continue to teach as an adjunct Associate Professor.

My dissertation, published March 14, 2014, by ProQuest, was entitled *Virtual Teams: Individual Perceptions of Effective Project Management That Contribute to a Collective Effort in Project Success* (Toler, 2014). Although I was hoping to condense my dissertation into a book, I found the requirements and formats for a book to be very different from what I had anticipated. Instead, I applied several alternative strategies in order to increase my opportunities to publish in the near-term and contribute the expert knowledge that I gained from my doctoral research as outlined below.

First, I extracted and adapted Chapters 3, 4, and 5 to meet the requirements of a peer-reviewed article, *Virtual Teams: Individual Perceptions of Effective Project Management That Contribute to a Collective Effort in Project Success*. This was published online by the Project Management Institute (PMI) and shared in PMI's virtual library.

Second, I extracted and adapted Chapters 3, 4, and 5 to meet the requirements to present an invited talk at the Institute of Nuclear Materials Management (INMM) 56th Annual Meeting, Indian Wells, California (2015). The full peer-reviewed article, *Developing Team Cohesiveness in a Virtual Environment*, was published in the Proceedings of the 56th INMM Annual Meeting (Toler, 2015).

Third, I extracted and adapted the literature review and Chapters 3, 4, and 5 contributing towards a book chapter 'Developing Project Team Cohesiveness in a Virtual Environment' in a handbook of research entitled *Strategic Management and Leadership for Systems Development in Virtual Spaces* published by IGI Global.

Fourth, I extracted sections and adapted them into weekly discussions and course guidance. These classes are at the undergraduate and graduate level, and I have been teaching them since 2014. The courses include Organisational Behaviour, Organisational Change Management, and Strategic Management for Organisations.

I did not expand on the original research itself. I found that the research fully covered the publication requirements, and in some cases exceeded it. I was able to publish soon after completing the dissertation, but if I were to publish using chapters from the dissertation currently, I would definitely expand upon the original research.

I did apply the findings in practical ways in my own professional career as a project manager. The research made me more aware of the special attention in communication efforts that is essential when working in a virtual environment. I also invite my students to apply the findings in their professional careers when leading virtual teams.

### *Don Dunn, PhD*

As a career-changer, post-PhD goals were to teach, though not to pursue a tenure-track position. In particular, I hoped to use the model I developed in my doctoral research to teach the next generation of managers to think and act ethically.

I wrote a book, *Designing Ethical Workplaces: The Moldable Model* (Dunn, 2016), designed for MBA-level students and corporate executives. For this, I was able to *extract* from the dissertation's *literature review* and from Chapters 4–5 using the *results* and the *recommendations* to build a theoretical foundation for the book and to practically demonstrate to students and executives how to use the *Moldable Model* to create an ethical workplace. The book is a *condensed* version of ethical leadership theory from the dissertation, coupled with the findings of the dissertation's research. These comprised a consistent, component model to design ethical workplaces, theoretically labeled *The Moldable Model*. *The Moldable Model's* fixed framework of a three-pronged system of role modeling, the context or reasons and results of why companies should be ethical, and holding employees accountable for company ethics became the main thrust of the book.

No, the book did not help in getting any work or job, at least to my knowledge; but yes, the book has added a great deal of respect as a faculty member. *Designing Ethical Workplaces* is used as a text for MBA and MS courses in ethics and in leadership. It is the basis for consulting with organisations about leadership, management, and ethics.

### Bessie Roan Bowser, PhD

My post-PhD career goals include a position in an online academic program where my experience as an online student will allow me to support other non-traditional students. Publications, including a tutorial, demonstrate that I have the knowledge and skills needed to serve as a faculty member.

When applying for positions, I have been asked to do a proposal that includes a conceptual or theoretical map. I have adapted the conceptual map from my dissertation to meet these application requirements.

The moment I hear calls for articles, chapters, or tutorials my thought process immediately focuses on my 2015 dissertation, *A grounded theory approach to creating a new model for understanding cultural adaptation of families in international assignments* (Bowser, 2015). I look at Chapter 5, which highlights results of the study, as well as theories and literature reviews that could support proposals.

For instance, I extracted the Family System Theory (FST) and prototype taxonomy of effective expatriation adaptation generated through grounded theory dissertation research. The theory was adapted for a book chapter, 'Banding organization, management, and leadership theories to identify managerial strategies' (Bowser, 2017).

I responded to a call to publish a tutorial; again, drawing on my dissertation, I extracted a theory central to the design of the dissertation research: cultural intelligence (CQ). I applied it to an introduction to leadership competencies for building global leaders (Bowser, 2018).

There are many more projects I can envision that build on my dissertation. I am planning to apply a great part of the dissertation into a book. To do so will require an expansion of the literature reviews to include research on ethics and sustainability.

### Joshua Fuehrer, PhD

My post-PhD career goals were to work as a management consultant. Rather than teach in an academic setting, I wanted to offer professional development for workers and managers in the workplace.

*Learning BPMN v2.0: A Practical Guide for Today's Adult Learners* (Fuehrer & Butchko, 2018) is central to my work as a management consultant. BPMN stands for Business Process Modeling and Notation. Clients like being able to use the book and related materials to implement approaches developed during consulting engagements.

To write the book, I expanded on my dissertation, conducting additional research using a mixed methods approach. First, quantitatively, I began assessing the models that were being delivered to various customers. I was examining the modeling errors that occurred both before and after individuals learned BPMN. Second, qualitatively, my co-author Joseph Butchko and I had been observing how learning occurred in various organisations. We engaged in conversations with individuals to understand their learning experiences with BPMN. The findings and data points from these informal interviews were used to expand on my dissertation findings. This was key for applying various learning modes and techniques throughout the book, really trying to create a set of meaningful learning experiences.

After consulting with various independent publishers, I chose an independent self-publishing service because the time it would take to get to market was faster, and they offered many key services needed to launch a great book. I struggled with the decision on whether to go down the route of conventional academic publishing, self-publishing, an indie publisher, or hybrid type publisher. After careful research, I quickly realised that self-publishing would require me to learn a lot of different skills to be successful. I think conventional commercial publishing is great, but being a new author with so much competition, it was an easy choice to go with an independent publishing company. This allowed us to access design and editorial services aimed to improve the quality of self-published books.

The book is promoted on active social media accounts, a blog, and guest posts on other blogs. On the book website, a forum allows readers to exchange ideas and a learning centre offers instructional videos, activities, and sample data for practice exercises.

## Chapter summary

The PIPs and students' stories illustrate the potential for strategic and thoughtful use of doctoral writing for *traditional and self-published books, articles, tutorials, and instructional materials.* While for some career paths, such as full-time academic positions, publications are needed before you can be considered as a candidate, for others publishing might evolve to complement career directions.

## Exercises and questions for reflection: finalise your publication strategy with short-term goals and a multi-year plan

You have completed a series of exercises throughout the book. Now it is time to pull your work together and generate a usable publication strategy and action plan. These final exercises will help you to build on what you have developed by studying Chapters 1–11, set your priorities, and determine how to move forward. Find

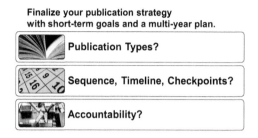

Finalize your publication strategy
with short-term goals and a multi-year plan.

| | |
|---|---|
| | **Publication Types?** |
| | **Sequence, Timeline, Checkpoints?** |
| | **Accountability?** |

*Figure 12.1* Finalise your publication strategy.

templates and worksheets for these exercises and related resources on the Routledge book website and www.path2publishing.com.

## Reflection

Now that you can see the contours of your publication strategy, how do you feel about this accomplishment? At the same time, what concerns do you have? What obstacles do you fear? Reflect on ways you can use your strengths, commitment to success, and your networks of friends and colleagues, to stay on track.

## Exercises

In the first exercise, you are asked to think through the strengths and weaknesses, opportunities and threats for each of your major publication goals. Be honest! If you are aware of weaknesses or threats that could jeopardise successful implementation of the plan, it is better to identify them now. This analysis should help you to think through all of the factors involved, and plan accordingly.


## Exercise 1: SWOT analysis of your publication strategy

Where do you stand? Complete a SWOT (strengths-weaknesses-opportunities-threats) analysis.

**Step 1:** Based on the strategy you have outlined, identify your strengths, weaknesses, opportunities, and threats. For example, Table 12.4 shows a sample SWOT analysis for a co-authored textbook.

**Step 2:** Look at the lists of strengths, weaknesses, opportunities, and threats. Note whether each item is related to one of the following factors (add additional categories as relevant):

a   Personal (e.g., time, confidence, skills)
b   Resources (e.g., funds, travel, hardware/ software)

*Table 12.4* Sample SWOT analysis

| Strengths | Weaknesses |
|---|---|
| • Recent data and up-to-date literature from the dissertation.<br>• According to feedback, I have a clear writing style. | • I have never written a book proposal, and do not have name recognition from prior publications. |
| **Opportunities** | **Threats** |
| • Topic is of current interest for undergraduate courses in my field.<br>• Co-author interested in writing a book together. | • There are already many textbooks on this topic on the market. |

c  Research/content (e.g., dated study, research applicable to narrow audience)

d  Partners, collaborators, assistance (e.g., co-authors, co-researchers, student assistants, or availability of other help)

e  Publisher/editor (e.g., finding the appropriate channel for publishing your work, making connections)

f  Audience/readers (e.g., marketing, promoting the book or article)

g  Societal/global (e.g., trends in the field, economic constraints, technological innovations)

h  Other

**Step 3: What have you learned?** Identify expert advice, skills, knowledge, or resources needed to achieve your goals.

### *Exercise 2: aligning career goals with your publication strategy*

Next, reflect on your career goals and identify the specific benefits associated with various types of publications. In earlier exercises you also reflected on your career goals. What, if anything, has changed since you started thinking about your publication and career aspirations? Have you uncovered new possibilities or gained the confidence to approach a goal you thought was out of reach?

In this exercise, think about your publication strategy and how each type of publication might help you to achieve your career goals. Change the columns to reflect the type(s) of career options you aspire to reach.

Table 12.5 offers an example.

The examples in the table are quite general. Using the blank Table 12.5 template on the book website, you have an opportunity to complete a similar table with more specific and personal examples. As always, feel free to add or remove rows or columns as necessary to make the table work best for you.

Table 12.5 Publication benefits

| | Benefits if you are: | | |
| Type of Publication | Employed in Academia | Employed Outside Academia | Consultant, Freelancer, Self-employed |
| --- | --- | --- | --- |
| **Case study** | Could be used as an instructional resource | Build your reputation as an expert | Could be used in a training activity |
| **Traditionally published book** | Book could be adopted as a text | You are regarded as an expert | Clients take you more seriously |
| **Self-published book** | Get your work in front of people very quickly | Build your reputation as an expert | Create a workbook you can use to complement training sessions |
| **Academic article in pay-walled peer-reviewed journal** | Counts towards tenure or promotion | Useful if you ever want to move into academia | Maximum respect from academia |

(Continued)

Table 12.5 Publication benefits (Continued)

| Type of Publication | Benefits if you are: | | |
| --- | --- | --- | --- |
| | Employed in Academia | Employed Outside Academia | Consultant, Freelancer, Self-employed |
| **Academic article in open-access journal** | Colleagues outside of academia can access the article | Students, clients can access the article | Gain some respect from academia |
| **Mainstream media piece** | Build your reputation as an expert, promote and sell your books | | Useful for promoting your consulting or training as well as any publications |
| **Social media piece** | Build your network, build your credibility, find new partners | | |
| **Podcast or audiobook** | Build your network, build your credibility | | |
| **Book chapter** | Good on CV | | |

## *Exercise 3: timeline and sequence of activities to achieve your goals*

These reflections and analyses bring you to the third exercise of this chapter: the timeline and sequence of activities.

In Exercise 2 you looked at your career goals and thought about how they relate to your publication goals. Now it is the time to:

- Create a timeline.
- Plan to start now with immediate steps.
- Identify milestones and checkpoints associated with each goal.
- Create an accountability system, so you keep moving forward.

If you want to work on publications based on the findings while the data and literature are fresh, note that point. If you now realise you need to contract with a cover designer or learn how to use a blogging platform, you can build the steps into this sequential plan. If the type of publication process involves periods of waiting while the peer review occurs, what could you be working on in the meantime?

This exercise is *absolutely key to your success*. It will help you to create a timetable to work from – the action plan to your strategy. Like your publication strategy itself, your timetable will need to be reviewed and revised at regular intervals, because external factors can cause changes in your professional or personal life, which you will need to take into account.

**Create a timeline.** First, take each of your planned publications and decide which year of your strategy it belongs to. Complete Table 12.6 below, using the lines that are relevant to you (i.e., if you have a two-year strategy, use only the first two lines). You can use the template on the Routledge website, or create your own.

*Get ready to start now!*

For the first six to twelve months, set yourself monthly tasks to complete. So, for example, if you are planning to work on a journal article first, you might need to start with background reading

*Table 12.6* Timeline

| Year | Monthly Targets | Skills, Steps, and Publications |
|------|-----------------|-------------------------------|
| **Year One** | | |
| **Year Two** | | |
| **Year Three** | | |
| **Year Four** | | |
| **Year Five** | | |

and note-taking on a specific topic. If your first publication is scheduled to be a blog post, you can probably dive straight into the writing. Once the bulk of your background work is done, you should be able to write a blog post in a week, a journal article in a month. If you're working on a longer piece, such as a full-length textbook or manual, set yourself a weekly word count and make sure you meet it. It's better to set a realistic word count you can manage rather than an over-ambitious one that will set you up to fail. For example, it should be possible to write 2,500 words per week (e.g., by writing 500 words or one typed page of A4, five days a week), which would yield an average-length journal article in two to three weeks or an average-length book in around seven months.

Outline your monthly tasks as far ahead as seems sensible to you. We recommend a minimum of six months and a maximum of 12. Don't forget to allow more leeway for times when writing may be more difficult, for example, times of year when your workload is particularly heavy, or when you'll be away on a family holiday. When you have completed the timeline exercise, you might want to revisit your publication strategy template and adjust as needed.

An important element of Exercise 3 is the system of accountability. You've made this concerted effort to develop your publication strategy – don't let those efforts go to waste! You know yourself: What do you need to make this work? Are you productive based on your own motivation and rewards? Or do you need external structure? Do you need a writing partner or a writing group to keep you going?

## Exercise 4: finalise your publication strategy

Many practical steps and personal commitments are needed to achieve your publication goals: It is beyond the scope of this book to address them all. But we know from experience that having – and using – a strategy will help you to advance towards the success you envision. In Exercise 4 you will finalise your very own publication strategy.

Start by answering these key questions:

- What steps do you need to add to your publication strategy, now that you have completed all of the book's exercises and activities?
- Who are the readers or users for each type of publication? How will you promote each type of publication to the target audience(s)? What budget or help will you need?
- Do you need to acquire new skills, such as how to set up a blog or how to write for a different

audience? Do you need to look for assistance, or contract for services you can't or don't want to do, such as an editor to review final copy for a self-published e-book, or graphic designer for a new logo, or a technology whiz who can help you set up for podcast recordings? If so, add the timeline and checkpoints for these items to your action plan.

Customise the outline suggestions below, depending on your writing projects and types of publications. Find the blank template in the Chapter 12 section on the book website. Copy and paste as appropriate into your working strategy document.

I Publication Type 1 (books, chapters, articles in peer-reviewed journals, articles in professional publications, blogs, or websites) _____

II Promotion _____
   a Target audience(s)
   b Goal (e.g., sales, text adoption)
   c Social media strategy
   d Mainstream media strategy
   e Paid advertising, boosting posts, etc.
   f Presentations (online or face-to-face)
   g Conference papers or other events
   h Book tour (online or face-to-face)

III Developing new skill and/ or contracting others to assist
   a Take a class
   b Find how-to resources
   c Find a coach or mentor
   d Find a partner whose skills complement
     mine
   e Budget to pay consultants (designers,
     editors, indexer, etc.)
IV Acquiring new technology tools or platforms
   a Criteria: What specific features are
     needed?
   b Free (potentially with advertising) versus
     paid
   c Budget

If you are planning to develop more than one
type of publication, answer the above questions
for each one.

## Good practice points

- Create a robust and detailed publication
  strategy.
- Use that strategy, together with knowledge of
  your personality and circumstances, to create an action plan to help you implement your
  strategy.
- Find a way to make yourself accountable.

# References

Bowser, B. R. (2015). *A grounded theory approach to creating a new model for understanding cultural adaptation of families in international assignments*. (PhD), Capella University, ProQuest.

Bowser, B. R. (2017). Banding organization, management, and leadership theories to identify managerial strategies. In L. L. West & A. Worthington (Eds.), *Handbook of research on emerging business models and managerial strategies in the non-profit sector* (pp. 126–151). Hershey, PA: IGI Global.

Bowser, B. R. (2018). *An introduction to leadership competencies for building global leaders*. Thousand Oaks, CA: SAGE Publications.

Dunn, D. (2016). *Designing ethical workplaces: The moldable model*. New York, NY: Business Expert Press.

Fuehrer, J., & Butchko, J. (2018). *Learning BPMN v2.0: A practical guide for today's adult learners*. Indie Books International.

Rallis, S. F., & Rossman, G. B. (2012). *The research journey: Introduction to inquiry*. New York, NY: Guilford Press.

Toler, L. (2014). *Virtual teams: Individual perceptions of effective project management that contribute to a collective effort in project success*. (PhD), Capella University, ProQuest.

Toler, L. (2015). *Developing team cohesiveness in a virtual environment*. Paper presented at the Institute of Nuclear Materials Management Indian Wells.

# Afterword

Congratulations! You have a publication strategy that is tailored to your unique situation and designed to help you achieve your career goals. You also have an action plan for implementing your publication strategy and a timeline with monthly tasks to complete. (If you don't have all of these, you might want to go back and complete the relevant exercises.)

The next step is to start work on your tasks for completion. As you do this, you will need to *use* your strategy and action plan. They are not documents to consign to a folder; they are for you to refer to and update regularly. Some people prefer to do this as they go along, while others like to make a regular commitment, say at the end or the beginning of each month. It doesn't matter how you work as long as your method is effective. However, you will need to take time now and again to step back from your tasks and give your strategy and action plan a thorough review. This is because circumstances change constantly and will inevitably have an impact on your plans.

Remember that rejection is a normal part of the publishing process. A rejection of your work is not a sign that you are a bad person or that your work is garbage. Your work may be rejected for any number of reasons other than its quality. For example, perhaps the editor doesn't think it fits their journal or series, or maybe they have just accepted something very similar, or they don't see your work as commercially viable for them. Also, rejection always comes with information – sometimes only a little, sometimes a lot – and that is invaluable information you probably couldn't get any other way. So, as the song says, pick yourself up, dust yourself down, and start all over again.

The more you write for publication, the easier it gets, which is not to say it ever gets truly easy. Writing is always intellectually demanding, but, as with any skill, regular practice leads to fluency. You may only want to publish here and there, or you may see writing as a regular part of your work. Either way, writing regularly will help you achieve your aims.

In this book we have given you the tools you need to achieve your publishing ambitions. All that remains is for us to wish you the very best of luck, as luck also has a part to play in publishing success. Though it is a truism that the harder you work, the luckier you get, this certainly applies to writing: the more you write, submit, rewrite, and resubmit, the more publications you will have to your name.

We'd love to hear about your progress on Twitter where we're @einterview and @DrHelenKara. Visit www.path2publishing.com for more resources about writing and publishing. Thank you for reading.

# Index